Springer Theses

Recognizing Outstanding Ph.D. Research

Aims and Scope

The series "Springer Theses" brings together a selection of the very best Ph.D. theses from around the world and across the physical sciences. Nominated and endorsed by two recognized specialists, each published volume has been selected for its scientific excellence and the high impact of its contents for the pertinent field of research. For greater accessibility to non-specialists, the published versions include an extended introduction, as well as a foreword by the student's supervisor explaining the special relevance of the work for the field. As a whole, the series will provide a valuable resource both for newcomers to the research fields described, and for other scientists seeking detailed background information on special questions. Finally, it provides an accredited documentation of the valuable contributions made by today's younger generation of scientists.

Theses are accepted into the series by invited nomination only and must fulfill all of the following criteria

- They must be written in good English.
- The topic should fall within the confines of Chemistry, Physics, Earth Sciences, Engineering and related interdisciplinary fields such as Materials, Nanoscience, Chemical Engineering, Complex Systems and Biophysics.
- The work reported in the thesis must represent a significant scientific advance.
- If the thesis includes previously published material, permission to reproduce this must be gained from the respective copyright holder.
- They must have been examined and passed during the 12 months prior to nomination.
- Each thesis should include a foreword by the supervisor outlining the significance of its content.
- The theses should have a clearly defined structure including an introduction accessible to scientists not expert in that particular field.

More information about this series at http://www.springer.com/series/8790

Alan Lewis

Spin Dynamics in Radical Pairs

Doctoral Thesis accepted by
the University of Oxford, UK

Springer

Author
Dr. Alan Lewis
The James Franck Institute
University of Chicago
Chicago, IL, USA

Supervisor
Prof. David Manolopoulos
Department of Chemistry
University of Oxford
Oxford, UK

ISSN 2190-5053 ISSN 2190-5061 (electronic)
Springer Theses
ISBN 978-3-030-00685-3 ISBN 978-3-030-00686-0 (eBook)
https://doi.org/10.1007/978-3-030-00686-0

Library of Congress Control Number: 2018954809

This Springer imprint is published by the registered company Springer Nature Switzerland AG
The registered company address is: Gewerbestrasse 11, 6330 Cham, Switzerland

Supervisor's Foreword

Organic radical pairs arise in a wide variety of contexts ranging from thin film physics and materials science to chemistry and biology. The Hamiltonians that govern the evolution of the electronic and nuclear spins in these radical pairs, and the equations that allow for the electron spin relaxation and radical pair recombination processes that compete with this coherent spin evolution, have the same form in all of these contexts. The resulting evolution equations are straightforward to write down, and it is also straightforward to derive expressions for all relevant experimental observables. However, these expressions are extremely expensive to evaluate on a computer for all but the smallest of radical pairs, because of the exponential scaling of quantum mechanics with system size. This makes exact quantum mechanical simulations of radical pairs containing more than a handful of hyperfine-coupled nuclear spins totally impractical using standard techniques.

In this thesis, Alan Lewis describes new quantum mechanical methods and semiclassical approximations which can be used to overcome this exponential scaling and make simulations of much larger radical pairs more practical. He then applies these techniques to a variety of problems that arise in disciplines ranging from materials science to biology. An exact quantum mechanical method that exploits the properties of spin coherent states is first applied to the problem of spin-dependent charge recombination along para-phenylene molecular wires, where it is used to extract spin-dependent (singlet and triplet) radical pair recombination rate constants from experimental data. A semiclassical theory based on the precession of classical electronic and nuclear spin vectors is then applied to two separate problems relating to avian magnetoreception: the simulation of a carotenoid-porphyrin-fullerene (CPF) radical pair that has recently been established as a "proof-of-principle" for the operation of a chemical compass, and simulations of the anisotropy of the singlet yield in a flavin-tryptophan radical pair in cryptochrome that has been suggested to play a role in the magnetic compass sense of migratory birds. Finally, a more primitive semiclassical theory due to Schulten and Wolynes is used to simulate the magnetoelectroluminescence and magnetoconductance of both deuterated and undeuterated DOO-PPV organic polymer light emitting diodes. Interesting physical insights are gained in all of these applications, none of which would have been possible using standard quantum mechanical techniques.

Much of the material presented in the thesis has now been published in scientific papers, including the material in the final chapter on triphasic behaviour in the time-dependent survival probability of the CPF radical pair (A. M. Lewis et al., J. Chem. Phys., 149, 034103, 2018). However, Alan's thesis does contain some details that have not been published elsewhere, including in particular his proof in Sect. 5.1.1 that the radical pair mechanism provides an inclination compass rather than a polarity compass (i.e. that the singlet yield of a radical pair recombination reaction is unchanged on reversing the direction of an applied magnetic field). It is well established from behavioural experiments that migratory birds have an inclination compass, and that this becomes inoperative in the absence of blue-green light. These two facts are among the strongest pieces of circumstantial evidence in favour of the (photochemically induced) radical pair mechanism of avian magnetoreception, so it is nice to see how the inclination compass arises. I am not aware of such a proof having been given before in the literature.

We are now continuing to investigate various aspects of radical pair spin dynamics in Oxford, using both the techniques that Alan describes in his thesis and new techniques that we have developed since he graduated. On the magnetoreception front, we have investigated the sensitivity of the directional information provided by the radical pair mechanism to the presence of weak radiofrequency magnetic fields, and the extent to which this directional information can be used to extract a compass bearing under low light conditions. We have also developed a new quantum mechanical method for studying radicals and radical pairs with very many hyperfine-coupled nuclear spins, and used it to investigate the hyperfine-induced decoherence of electron spins in semiconducting quantum dots. All of this work has been or will soon be published in the open literature.

There are clearly many different physical, chemical and biological problems that one can solve when one knows how to simulate the spin dynamics of radical pairs. We now have the tools to do this for arbitrarily large radicals, thanks in part to the developments described in this thesis. It will be interesting to see how much more can be done with these tools in the future.

Oxford, UK Prof. David Manolopoulos
July 2018

Abstract

The coherent spin dynamics of radical pairs play a crucial role in their reactions, which consequently cannot be described by a simple kinetic scheme. Instead, simulations of the spin dynamics are required in order to predict the rate and outcome of radical pair reactions, and especially their response to the application of a magnetic field. Unfortunately, the number of spin states of the radical pair increases exponentially with the number of nuclear spins, making deterministic quantum mechanical simulations of realistic radical pairs difficult.

To overcome this difficulty, this thesis begins by presenting an efficient stochastic quantum mechanical method capable of describing a radical pair with as many as 20 nuclear spins, which we use to analyse spin-dependent charge recombination rates along molecular wires. This enables us to identify the mechanism of charge recombination of both the singlet and triplet states of the wire by determining their relative contributions to the overall recombination rate.

We then derive an approximate semiclassical theory which allows to treat the spin dynamics of much larger radical pairs, since the time required for a semiclassical calculation scales linearly with the number of nuclear spins, rather than exponentially. Using this method, we reproduce the results of the first experiments to show that the outcome of a radical pair reaction may be influenced by an Earth-strength magnetic field, and calculate the anisotropy in the singlet recombination yield of the radical pair thought to be responsible for avian magnetoreception.

We show that our semiclassical theory reduces to the earlier Schulten–Wolynes theory under two additional approximations, and use this simpler theory to reveal that singlet-triplet dephasing plays an important role in the spin dynamics of polaron pairs in the semiconducting polymer layer of organic light emitting diodes. We derive a new expression which relates the magnetic field dependence of the electroluminescence and conductance observed in these materials to the singlet yield of the radical pair recombination reaction, which we confirm produces better agreement with experimental data than the relationships used previously.

Parts of this thesis have been published in the following journal articles:

Manolopoulos, D. E. & Hore, P. J. An improved semiclassical theory of radical pair recombination reactions. *Journal of Chemical Physics* **139**, 124106 (2013).

Lewis, A. M., Manolopoulos, D. E. & Hore, P. J. Asymmetric recombination and electron spin relaxation in the semiclassical theory of radical pair reactions. *Journal of Chemical Physics* **141**, 044111 (2014).

Lawrence, J. E., Lewis, A. M., Manolopoulos, D. E. & Hore, P. J. Magnetoelectroluminescence in organic light-emitting diodes. *Journal of Chemical Physics* **144**, 214109 (2016).

Lewis, A. M., Fay, T. P. & Manolopoulos, D. E. An efficient quantum mechanical method for radical pair recombination reactions. *Journal of Chemical Physics* **145**, 244101 (2016).

Hiscock, H. G. et al. The quantum needle of the avian magnetic compass. *Proceedings of the National Academy of Sciences of the United States of America* **113**, 201600341 (2016).

Fay, T. P., Lewis, A. M. & Manolopoulos, D. E. Spin-dependent charge recombination along para-phenylene molecular wires. *Journal of Chemical Physics* **147**, 064107 (2017).

Lewis, A. M. et al. On the low magnetic field effect in radical pair reactions. *Journal of Chemical Physics* **149**, 034103 (2018).

Acknowledgements

I must begin by thanking my supervisor, Prof. David Manolopoulos, for all of the time he has spent teaching me how and why to conduct research. Thanks to Prof. Peter Hore for all of his expert advice on spin dynamics, and for putting us on the right track time after time. Thanks to Joseph Lawrence and Tom Fay for being excellent colleagues and for a whole host of bright ideas about oLEDs and molecular wires. I would like to thank Susannah Worster for her helpful explanations of Redfield theory and other descriptions of relaxation, and Hamish Hiscock for his comments on Chap. 5, as well as their invaluable help in tackling spin dynamics with me. Huge thanks must go to my parents, Joy and Andy, and my brother Mark, for encouraging, challenging and supporting me throughout my academic career. Thank you to my loving wife Hannah, who has taken a great interest in my research, patiently listened to all of my jelly baby-based explanations of quantum chemistry, and selflessly given her time to help me complete this thesis. Finally, I thank Jesus; the Creator, Author and Perfector, from whom comes all of my ability and desire to study His world.

Contents

List of Figures

List of Tables

Chapter 1
Introduction

Radical pair reactions play an important role in a wide range of biological and technological processes. The overall rates and product yields of these reactions are strongly dependent on the spin dynamics of the radical pair, despite the fact that the magnetic interactions which produce these dynamics are far weaker than the thermal energy at room temperature. In this thesis we will outline the quantum mechanics which describes these radical pair reactions, before deriving semiclassical approximations which allow the computation of the ensemble averages of observables in realistic radical pairs. We will then apply these methods to three different problems in order to gain an insight into the physical processes which affect the rate and outcome of these radical pair reactions.

In this chapter we will first introduce the property of spin and the quantum mechanical description of two correlated spins. We will then outline the radical pair mechanism, which demonstrates how the spin dynamics of the unpaired electrons in a radical pair affect the outcome of its reaction. A qualitative explanation of how radical pair reactions are influenced by the application of a magnetic field is then given, before we briefly discuss the different radical pair systems which we will consider in this thesis. Finally, we set out the structure of the remainder of this thesis.

1.1 Spin

Spin is a fundamental property of electrons and nuclei which emerges from solving the relativistic Schrödinger equation [1]. It is described as an intrinsic angular momentum, and has no classical analogue [2]. The spin angular momentum is described by two quantum numbers: the spin quantum number, S, and the spin projection quantum number, M_S. These are defined in reference to the eigenstates of the quantum mechanical operator for the square magnitude of the angular momentum, $\hat{S}^2$, and the operator for the projection of the spin on the z axis, $\hat{S}_z$ [3]:

A. Lewis, *Spin Dynamics in Radical Pairs*, Springer Theses,
https://doi.org/10.1007/978-3-030-00686-0_1

$$\hat{S}^2 \,|S, M_S\rangle = \hbar^2 S(S+1)\,|S, M_S\rangle\,,$$
$$\hat{S}_z \,|S, M_S\rangle = \hbar M_S \,|S, M_S\rangle\,. \tag{1.1}$$

S can take any integer or half-integer value; M_S can take any value between S and $-S$ in integer steps. In the case of an electron, $S = 1/2$, so there are two electronic spin states with $M_S = \pm 1/2$. The following compact notation is commonly used for these states:

$$|1/2, 1/2\rangle = |\alpha\rangle\,, \quad |1/2, -1/2\rangle = |\beta\rangle\,. \tag{1.2}$$

Since spin are angular momenta, they can couple [2]. For example, two electron spins, $\mathbf{S}_1$ and $\mathbf{S}_2$, can couple to produce a total angular momentum $\mathbf{S}$, which may have a spin quantum number of either $S = 1$ or $S = 0$. In the former case, there are three possible values of M_S (0, ± 1); the three corresponding states are referred to as the triplet states [4]. When $S = 0$ there is only one possible value of M_S (0); this state is called the singlet state. These coupled states are related to the uncoupled electron spin states as follows [5]:

$$\begin{aligned} |1, 1\rangle &= |\alpha\rangle_1 \otimes |\alpha\rangle_2\,, \\ |1, 0\rangle &= \frac{1}{\sqrt{2}}\Big(|\alpha\rangle_1 \otimes |\beta\rangle_2 + |\beta\rangle_1 \otimes |\alpha\rangle_2 \Big), \\ |1, -1\rangle &= |\beta\rangle_1 \otimes |\beta\rangle_2\,, \\ |0, 0\rangle &= \frac{1}{\sqrt{2}}\Big(|\alpha\rangle_1 \otimes |\beta\rangle_2 - |\beta\rangle_1 \otimes |\alpha\rangle_2 \Big). \end{aligned} \tag{1.3}$$

These coupled spin states are commonly referred to using the shorthand

$$|1, 1\rangle = |\mathrm{T}_+\rangle\,, \quad |1, 0\rangle = |\mathrm{T}_0\rangle\,, \quad |1, -1\rangle = |\mathrm{T}_-\rangle\,, \quad |0, 0\rangle = |\mathrm{S}\rangle\,. \tag{1.4}$$

The energies of these states depend principally on two factors: the strength of the applied magnetic field, B [2], and the exchange coupling, J [6, 7]. The origins of these interactions will be discussed in detail in Sect. 2.1; for now, a qualitative sketch of their effects will suffice. Figure 1.1 shows how the relative energies of the singlet and triplet states vary as a function of the applied field strength B in the case of a positive exchange coupling constant J. The exchange coupling lifts the degeneracy of the singlet and triplet states in the absence of a field, and when a magnetic field is applied the degeneracy of the three triplet states is also broken, as the $|1, 1\rangle$ and $|1, -1\rangle$ states increase and decrease in energy respectively [8, 9]. This symmetry breaking and separation of energy levels gives rise to observable magnetic field effects, discussed in more detail in Sect. 1.3.

Many nuclei also possess spin angular momentum, which is denoted with the quantum number I to differentiate it from the spin of an electron. Of particular note are the ^{1}H and ^{14}N nuclei, which have spin numbers $I = 1/2$ and $I = 1$ respectively [10], as they are ubiquitous in the organic molecules considered in this thesis. Both

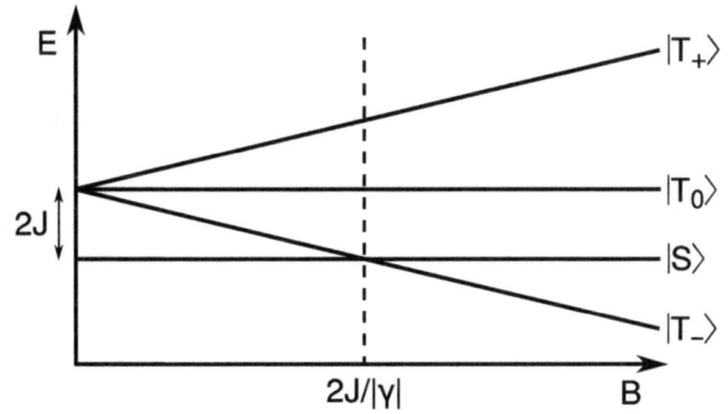

Fig. 1.1 The relative energies of the singlet and triplet states as a function of the applied field strength B. γ is the gyromagnetic ratio of the electron. In the absence of a magnetic field the energies of the singlet and triplet states are split by the exchange coupling of magnitude J. We have assumed here that the exchange coupling is antiferromagnetic ($J > 0$); if it were ferromagnetic the singlet state would be higher in energy than the triplet state in the absence of a magnetic field

^{12}C and ^{16}O nuclei have $I = 0$, so they can be ignored in the context of spin dynamics [10]. ^{13}C has $I = 1$, but is present at only 1% abundance [11], and so is neglected in this thesis.

In systems where spin-orbit coupling is negligible, chemical reactions and photochemical processes conserve the total electronic spin angular momentum [12–14]. Radical pair reactions are a particular class of reactions where this conservation of momentum plays a crucial role. Radical pairs contain two unpaired electrons whose spins are coupled, and so exist in either the singlet or triplet state. This electronic spin state is often critical in determining the product yields and rate of reaction, even though the energy differences between spin states are very small relative to the thermal energy [13, 15]. These reactions are described in detail in Sect. 1.2, and will be the focus of this thesis.

1.2 Radical Pair Reactions

A general reaction scheme for radical pairs based on the mechanism proposed independently by Closs [16] and Kaptein and Oosterhoof [17] is shown in Fig. 1.2. Photoexcitation of a target molecule AB is followed by rapid charge separation to form a radical pair, typically in the singlet state provided that spin-orbit coupling is negligible. If the electron spins interact with the nuclear spins in the radical, the singlet state is not stationary – the radical pair will undergo intersystem crossing to form the triplet state. The singlet and triplet states can each recombine, in general at different rates and to form different products. Since the intersystem crossing is a coherent process, this reaction scheme cannot be described by a simple kinetic model. Therefore, the detailed spin dynamics of the radical pair must be simulated in order to

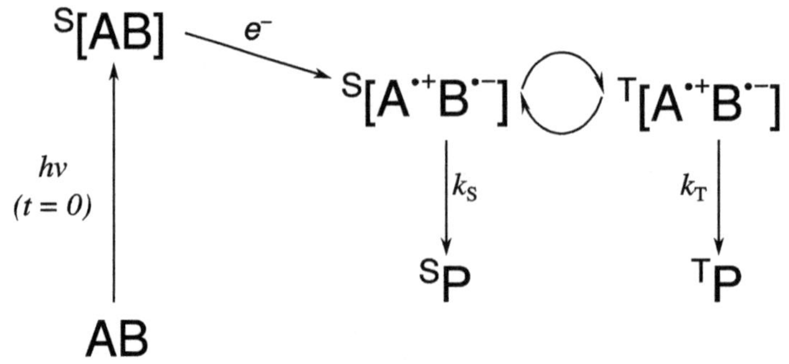

Fig. 1.2 An idealised radical pair recombination reaction. The superscript S and T indicate singlet and triplet states, respectively. The singlet product ^{S}P may or may not be the ground state, and the charge separation process may occur in multiple steps rather than the single step shown here

predict the outcome of its reaction. The coherent intersystem crossing is influenced by an external magnetic field, and so the product yields and radical pair lifetimes may be modified by the application of a field – these effects are described generally as magnetic field effects (MFEs) [9, 14, 18, 19].

The strength of the hyperfine interactions in a radical can be described by a single parameter, the hyperfine field [20, 21]:

$$B_{\text{hyp}} = \sqrt{\sum_k (a_k/|\gamma|)^2 \, I_k(I_k + 1)}. \tag{1.5}$$

Here the sum is over all of the nuclear spins in the radical, a_k is the isotropic hyperfine constant associated with nuclear spin k, I_k is its spin quantum number, and γ is the gyromagnetic ratio of the electron. This approximate measure of the hyperfine interactions is directly comparable to the strength of an external magnetic field B, and can be used to help explain the form of the magnetic field effects described in Sect. 1.3.

Schulten and Wolynes' semiclassical treatment of radical pairs [20], discussed in detail in Sect. 3.2, provides an intuitive way to understand why the hyperfine interactions result in intersystem crossing between the singlet and triplet electron spin states. Within a vector model, electron spins are thought of as precessing around a magnetic field, with the precession frequency determined by the strength of the field. If two correlated spins precess at the same frequency around the same axis, their relative orientations will remain constant and their spin state will not change. However, in the presence of hyperfine interactions, each electron spin can be thought of as precessing around an effective magnetic field

$$\mathbf{B}_{eff,i} = \mathbf{B} + \mathbf{B}_{\text{hyp},i}, \tag{1.6}$$

where **B** is the applied magnetic field. Even though **B** will be the same for each radical in the pair, $\mathbf{B}_{\mathrm{hyp},i}$ will differ, leading to different rates of precession around different axes and a change in the relative orientation of the two spins, and hence a change in the spin state of the radical pair. Note that in this simplified picture, the precession of the nuclear spins is ignored, the consequences of which are discussed in Sect. 3.3. Nevertheless, this provides a helpful qualitative picture of hyperfine-mediated intersystem crossing in radical pairs.

For each particular case we consider in this thesis, it will be necessary to discuss some variation of this general scheme. These differences will be covered in detail in the relevant chapters in the thesis, but Fig. 1.2 remains a good general guide to any radical pair reaction.

1.3 Magnetic Field Effects

The application of a magnetic field affects the spin dynamics of a radical pair, and in general changes its product yields and lifetime. The precise effect produced will depend on the initial spin state of the radical pair, the observable of interest, and the relative magnitudes of the different magnetic interactions within the radical pair. However, the magnetic field effects observed in radical pairs in which the dominant interactions are the Zeeman interaction, hyperfine interactions, and an exchange coupling can be categorised by considering three limiting cases. These are: the high field effect, the application a magnetic field much larger than the hyperfine fields; the low field effect, the application of a very weak but non-zero magnetic field; and the resonance effect, the response of a radical pair with an exchange coupling larger than its hyperfine fields to a magnetic field. These will be illustrated by considering the dependence of the singlet yield of a singlet-born radical pair on the applied magnetic field strength, B.

1.3.1 The High Field Effect

In general, for the application of a magnetic field to affect the outcome of a radical pair reaction, it must change the rate of interconversion between the spin states of the radical pair. In this thesis, we are concerned with intersystem crossing induced by the hyperfine interactions between the electron and nuclear spins in each radical. The most obvious way in which a magnetic field will affect this process is through the Zeeman interaction, which separates the energies of the $|\mathrm{T}_{\pm}\rangle$ states from that of the $|\mathrm{S}\rangle$ state, as shown in Fig. 1.1. This significantly reduces the efficiency of any interconversion between the singlet and triplet states of the radical pair. Therefore, a high field effect (HFE) is observed when the external magnetic field is stronger than both the hyperfine interactions and the exchange coupling, $B \gg B_{\mathrm{hyp}}, |2J/\gamma|$ [22, 23].

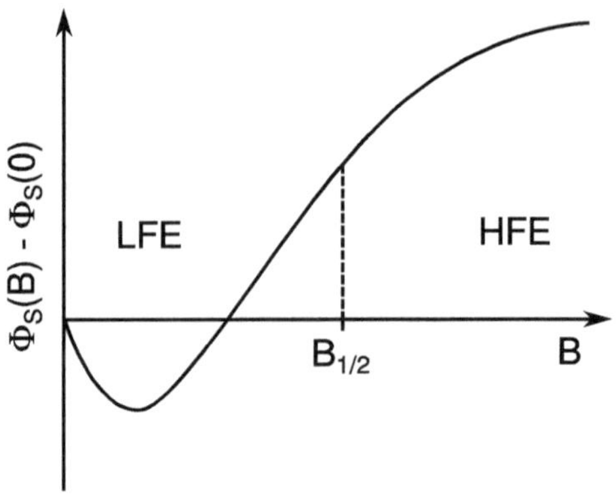

Fig. 1.3 An illustration of the low field effect (LFE) and high field effect (HFE) on the singlet yield of a singlet-born radical pair

For a singlet-born radical pair, the HFE results in an increase in the singlet yield, Φ_S, relative to the zero-field singlet yield, as the applied field strength is increased [24, 25]. This will eventually saturate once the magnetic field is large enough that essentially all $|S\rangle \leftrightarrow |T_\pm\rangle$ interconversion is suppressed. This dependence is shown qualitatively in Fig. 1.3 [21]. The HFE is characterised by the parameter $B_{1/2}$, the magnetic field strength at which the singlet yield is halfway between its value in the absence of a magnetic field and its value at saturation. $B_{1/2}$ clearly depends on the strength of the hyperfine interactions in each radical [23, 26], since saturation will only occur once the intersystem crossing they induce has been inhibited.

Weller et al. obtained an empirical relationship between $B_{1/2}$ and the weighted average of the effective hyperfine field of each radical [21],

$$\begin{aligned} B_{1/2} &= \frac{B_{\text{hyp},1}}{1/2\left(B_{\text{hyp},1} + B_{\text{hyp},2}\right)} B_{\text{hyp},1} + \frac{B_{\text{hyp},2}}{1/2\left(B_{\text{hyp},1} + B_{\text{hyp},2}\right)} B_{\text{hyp},2} \\ &= 2\frac{B_{\text{hyp},1}^2 + B_{\text{hyp},2}^2}{B_{\text{hyp},1} + B_{\text{hyp},2}}, \end{aligned} \tag{1.7}$$

which they showed agrees well with certain experimental values of $B_{1/2}$. However, we have found that in radical pairs where the hyperfine fields of the two radicals are very different, the Weller equation performs less well. Figure 1.4 plots the value of $B_{1/2}$ obtained from our simulations and that predicted by the Weller equation against the hyperfine field of one of the radicals in the pair, while the hyperfine field of the other radical is held constant at 1 mT. The simulations were performed using the Schulten-Wolynes method described in Sect. 3.2, considering only the Zeeman and hyperfine interactions. While the Weller equation is qualitatively accurate, and shows good agreement with the simulations when the hyperfine fields of the two radicals are similar, it becomes becomes much less accurate as the ratio $B_{\text{hyp},1}/B_{\text{hyp},2}$ increases.

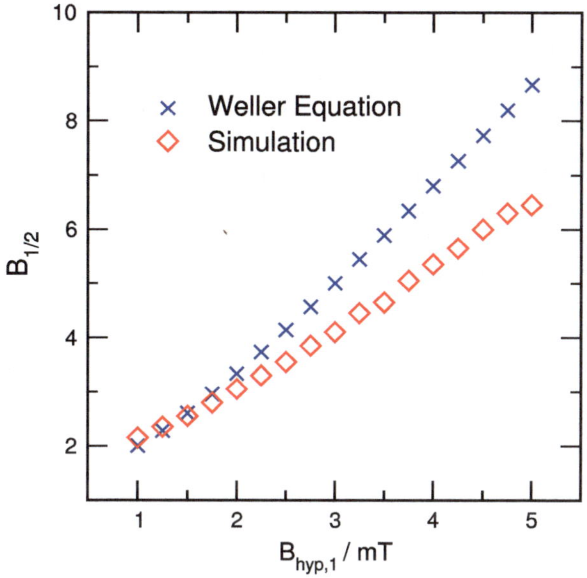

Fig. 1.4 $B_{1/2}$ as predicted by the Weller equation, Eq. (1.7), and extracted from simulations, as a function of the hyperfine field of radical 1. In each case the hyperfine field of the second radical was 1 mT. The simulations were performed using Schulten-Wolynes theory, which is described in Sect. 3.2, with the recombination rates of both the singlet and triplet radical pair chosen to be 10^6 s^{-1}. The simulated $B_{1/2}$ does not depend strongly of the choice of rate constants

1.3.2 The Low Field Effect

It is less obvious how the application of a magnetic field weaker than the hyperfine interactions, $B \ll B_{\text{hyp}}$, will affect the singlet-triplet intersystem crossing. Nevertheless, a low field effect (LFE) is observed experimentally [27–29] and predicted theoretically [25, 30], and serves to increase the rate of interconversion between singlet and triplet states, rather than decreasing it. This effect was originally identified with a change in which quantities are conserved upon the application of a magnetic field [18]. In the absence of a magnetic field, all three components of the total angular momentum of the radical pair are conserved; in the presence of a magnetic field only the projection on the field axis is conserved, allowing more conversion between the singlet and triplet states.

Timmel et al. suggested that the LFE could be understood as the result of symmetry-breaking [25]. In the absence of a magnetic field, the three electronic triplet states are degenerate, but applying a magnetic field breaks the symmetry of the system. This lifts the degeneracy between some of the eigenstates of the radical pair, causing the coherences between those two states to oscillate. If the two eigenstates in question both have some singlet character (that is, $\langle m|\hat{P}_{\text{S}}|m\rangle \neq 0$ for both eigenstates $|m\rangle$), then this change in the relative phase of the two eigenstates changes the probability of finding the radical pair in the singlet state. In other words, the application of a magnetic field breaks the zero-field symmetry of the radical pair and allows additional pathways for interconversion between the singlet and triplet electronic states. In the case of a singlet-born radical pair, this leads to a decrease in the singlet yield at low fields relative to the singlet yield in the absence of a field [25], as shown qualitatively in Fig. 1.6.

In order for a significant low field effect to be observed, the lifetime of the radical pair τ must be long enough and the relaxation rate k_R slow enough for intersystem crossing to take place. Since the rate of the additional interconversion induced by the low field is related to the Larmor frequency of the electrons in the radical pair, $\omega = -\gamma B$, one condition for observing a low field effect is $\tau^{-1}, k_R < \omega$ [25]. In addition, electron spin coupling will lower the degeneracy of the radical pair in the absence of a magnetic field, thereby reducing or even completely eliminating the low field effect in these radical pair reactions [25].

The maximum possible low field effect of a singlet-born radical pair is observed in the limit of extremely slow radical pair recombination, since this provides the most opportunity for conversion of the singlet state to the triplet state. In this limit, it is found that the magnitude of the low field effect is given by [31]

$$\frac{\Phi_S^0 - 1/4}{\Phi_S^{LF} - 1/4} = 3, \tag{1.8}$$

where Φ_S^0 is the singlet yield in the absence of a magnetic field and Φ_S^{LF} is the singlet yield in a low field. This can be understood using a simple vector model [31], illustrated in Fig. 1.5. In the absence of an applied magnetic field, the vector corresponding to the electron spin on radical i, $\mathbf{S}_i$, and the vector corresponding to a nuclear spin on that radical, $\mathbf{I}_i$, will precess around the resultant of their two vectors, $\mathbf{J}_i$. Within the vector model, the singlet yield is defined as

$$\Phi_S^0 = \frac{1}{4} - \overline{\langle \mathbf{S}_1 \cdot \mathbf{S}_2 \rangle} = \frac{1}{4} - \langle \mathbf{F}_1 \cdot \mathbf{F}_2 \rangle, \tag{1.9}$$

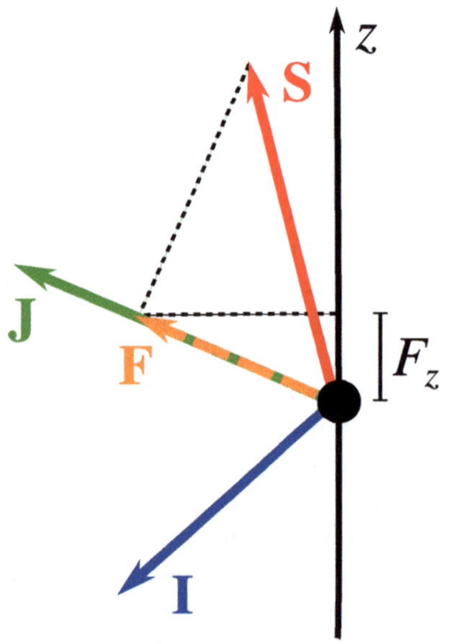

Fig. 1.5 A diagram illustrating the justification of Eq. (1.8) using the vector model. The electron spin vector **S** is shown in red, the nuclear spin vector **I** is shown in blue, the resultant of those vectors **J** is shown in green, and the projection of **S** onto that resultant is **F**, shown in orange. The projection of **F** onto the z axis is also indicated

where the brackets indicate the average over the different combinations of the resultant vectors $\mathbf{J}_1$ and $\mathbf{J}_2$, the overbar indicates a time average, and $\mathbf{F}_i$ is the projection of $\mathbf{S}_i$ onto $\mathbf{J}_i$. In the limit of zero field, the system is isotropic, so

$$\langle F_{1,x} F_{2,x} \rangle = \langle F_{1,y} F_{2,y} \rangle = \langle F_{1,z} F_{2,z} \rangle \tag{1.10}$$

and

$$\Phi_S^0 = \frac{1}{4} - 3 \langle F_{1,z} F_{2,z} \rangle. \tag{1.11}$$

However, when a weak magnetic field is applied, $\mathbf{J}_1$ and $\mathbf{J}_2$ precess around the z axis, so only the z components of $\mathbf{F}_1$ and $\mathbf{F}_2$ remain independent of time, and the singlet yield is given by

$$\Phi_S^{LF} = \frac{1}{4} - \overline{\langle \mathbf{S}_1 \cdot \mathbf{S}_2 \rangle} = \frac{1}{4} - \langle F_{1,z} F_{2,z} \rangle. \tag{1.12}$$

Combining Eqs. (1.11) and (1.12) we find an expression for the limiting value of the magnitude of the low field effect, Eq. (1.8).

1.3.3 Resonance Effect

Resonance effects are observed when the exchange coupling is large compared to the hyperfine field, $|2J/\gamma| > B_{\mathrm{hyp}}$. In these systems, singlet-triplet interconversion increases to a maximum as the magnetic field strength increases, before falling below its zero-field rate due to the normal high field effect [8, 32, 33]. This is qualitatively similar to the low field effect, but is observed for much larger applied fields, where $B > B_{\mathrm{hyp}}$. In this case, the energy of the singlet state is well separated from the energy of the triplet states in the absence of a field, resulting in little intersystem crossing. However, as the applied field strength is increased towards $|2J/\gamma|$, the energy of one of the triplet states approaches that of the singlet state and the rate of interconversion increases. This reaches a maximum at $B = |2J/\gamma|$ – the resonance condition, indicated in Fig. 1.1 [8, 9, 34]. As B is increased further still, the energy of two states separates again, and intersystem crossing is suppressed. Therefore, a minimum in the singlet yield of a singlet-born radical pair is observed when this condition is satisfied, as illustrated in Fig. 1.6.

The width of the resonance peak, denoted ΔB in Fig. 1.6, is usually related to the strength of the hyperfine interactions of the radical pair [21, 34, 35]. The larger B_{hyp} is, the less precisely the applied field strength must match the exchange coupling for interconversion to occur, and so the range of applied field strengths at which significant intersystem crossing can take place becomes wider. ΔB is also affected by the lifetimes of the radical pairs. As the lifetime of either the singlet or triplet state is reduced, its energy levels are broadened, producing a non-zero density of states at the energy of the other state over a wider range of magnetic field strengths around the resonance at $B = |2J/\gamma|$ [36]. This is discussed in more detail in Sect. 4.4.1.

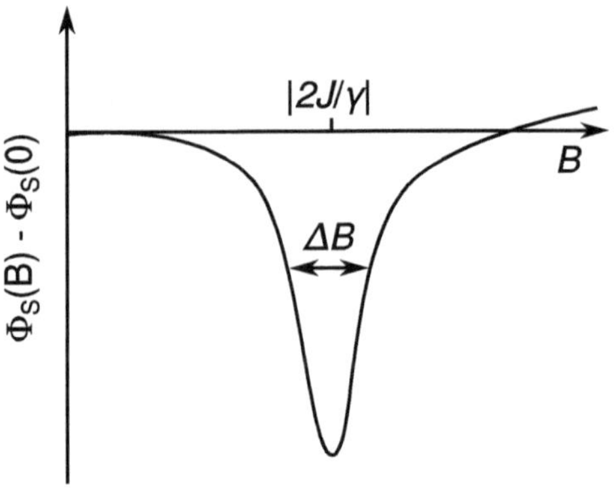

Fig. 1.6 An illustration of the resonance effect on the singlet yield of a singlet-born radical pair

Finally, this discussion has focussed on a single molecule with a well defined value of J. If, however, we consider an ensemble of flexible molecules, there will be a range of inter-radical separations and hence a distribution of values of J within an individual sample [37, 38]. In this case, the width ΔB observed in an experiment may arise from the superposition of different resonance effects in molecules with different values of J [39].

1.4 Applications

Our goal in this thesis will be firstly to develop both quantum mechanical and semi-classical descriptions of radical pair reactions, and then to use these methods to gain insight into real problems. We will now briefly introduce the specific examples of radical pair reactions to which we will apply the theory developed in this thesis.

1.4.1 Molecular Wires

Molecular wires are designed to imitate the long-range transport of electrons from chromophores to reaction centres observed in photosynthetic organisms [39–42]. This feature is extremely desirable in chemical solar energy conversion systems, which aim to mimic the behaviour of photosynthetic light harvesting complexes [39, 43], as well as in molecular electronics and nanotechnology [44, 45]. In order to synthesise molecular wires optimised for efficient charge transport, it is vital to understand the mechanisms by which electron transfer along them occurs.

There is a body of experimental evidence which shows that the distance dependence of the rate of charge recombination along molecular wires varies significantly between different molecular wires [46–49], and in some cases even changes as the length of the wire is increased [50]. This suggests that there are different mechanisms of electron transfer which operate in molecular wires. A better understanding of these mechanisms can be gleaned from the magnetic field dependence of the triplet and radical pair yields of charge-separated molecular wires, which have been measured in a number of different systems [34, 39, 50–52]. These experiments have been used to obtain the exchange coupling constant between the two unpaired electrons in the molecular wires, as well as allowing us to determine the relative contribution of the singlet and triplet recombination pathways to the overall rate of charge recombination, which depends on the spin dynamics of the radical pair.

1.4.2 Avian Magnetoreception

It has been well known for a long time that many organisms are able to navigate using the Earth's magnetic field [53–58]. More recently, radical pair reactions have been proposed as the mechanism by which the Earth's very weak magnetic field might be detected in biological systems [18, 59, 60]. Much of the biological research on magnetoreception has focussed on migratory birds [61–65], which will be the focus of Chap. 5 of this thesis.

The first experimental demonstration that a radical pair reaction could act as a magnetoreceptor in an Earth-strength magnetic field was performed by Maeda et al. in 2008 [66]. However, these proof-of-principle experiments were performed using a carotenoid-porphyrin-fullerene triad optimised to detect a low field effect, rather than the cryptochrome-based radical pair thought to be responsible for magnetoreception in migratory birds [14, 15, 60, 67, 68]. The cryptochromes are photoreceptors found in both plants and animals which are involved in regulating the circadian clock [69]. and are found in the retinas of several species of bird which exhibit a magnetic compass sense [70]. The anisotropic response of the cryptochrome radical pair to magnetic fields which is required for them to function as magnetoreceptors has been difficult to measure experimentally. Instead, much theoretical work has been done to try to assess this, in an effort to ascertain the plausibility of the radical pair mechanism of magnetoreception [71–75].

1.4.3 Magnetoelectroluminescence

Organic light emitting diodes (oLEDs) are rapidly becoming commonplace in a variety of commercial applications [76–78], and offer several advantages over similar inorganic devices [79–81]. Despite this, there is still significant debate about the factors which govern the dynamics of the charge carriers in the electroluminescent semiconducting polymer layer of oLEDs [82–88]. The observation of a magnetic field effect on the electroluminescence of several semiconducting conjugated polymers, and especially an isotope effect on that MFE, has provided convincing evidence for the involvement of the radical pair mechanism in electroluminescence [80, 87,

89–92]. However, previous simulations of the magnetoelectroluminescence have all failed to quantitatively reproduce the experimental observations [86, 87, 89]. This suggests that the existing theory has neglected a magnetic field dependent aspect of the mechanism of electroluminescence. Furthermore, while isotopic substitution experiments have conclusively demonstrated the involvement of hyperfine interactions in the mechanism of magnetoelectroluminescence [86, 90], there may well be other physical effects which influence the spin dynamics of the radical pairs in oLEDs.

1.5 Outline of Thesis

The remainder of this thesis will be organised as follows. Chapter 2 outlines the quantum mechanical description of radical pair reactions. It introduces the Hamiltonian which governs the spin evolution of radical pairs, and briefly sketches the origin of each term which appears. It then defines the Haberkorn recombination operator used to account for the reaction of these radical pairs, and justifies its use rather than a recently proposed alternative. With these operators in place, it is possible to find an expression for the ensemble average of an observable of a radical pair reaction, which we may use to make comparisons between simulations and experimental results. These ensemble averages may be evaluated more efficiently when formulated in the coherent spin state basis of the nuclear spins of the radical pair. The performance of this method is then assessed. Chapter 2 concludes with a discussion of two different mechanisms by which the electron spins in radical pairs can relax, and the difficulty of including these effects in quantum mechanical calculations.

Chapter 3 is devoted to the semiclassical treatment of radical pair reactions. It opens by explaining why these approximate methods are often necessary to simulate real radical pair systems, before describing two of these methods in detail: the semiclassical (SC) theory, and the earlier and simpler Shulten-Wolynes (SW) theory. Both involve treating the electron and nuclear spins as classical vectors, with SW theory making additional approximations about the dynamics of the nuclear spin vectors. It then compares the performance of both of these methods with exact quantum mechanical results for three model systems, revealing that each method has a range of applicability in which it is both useful and reasonably reliable. Finally, it demonstrates how electron spin relaxation may be treated straightforwardly within the semiclassical theories, in contrast to the quantum mechanical description of radical pair reactions.

Chapter 4 begins by describing the two limiting mechanisms by which charge recombination along molecular wires may occur, and introduces the $PTZ^{\bullet+}$–Ph_n–$PDI^{\bullet-}$ molecular wires whose magnetic field effects we will reproduce. It then describes the parameters which define the Hamiltonian of this molecular wire, as well as the constraints on its recombination rates provided by experimental measurements. The results of our simulations are then presented, with a background contribution to the triplet yield proposed to explain the magnetic field dependence of the triplet yields observed experimentally. The chapter concludes by discussing three interesting features of our simulations: the width of the resonance peaks seen in the magnetic field effect on the triplet yield of the recombination reaction, the rates of recombination of the singlet and triplet radical pairs, and the origin of the background contribution to the triplet yield.

Chapter 5 starts by describing the experimental evidence for the radical pair mechanism of magnetoreception, and provides details of the cryptochrome radical pair proposed as the magnetoreceptor in birds. It then demonstrates that the results of the proof-of-principle experiments of Maeda et al. [66] can be reproduced using the SC theory, and that these simulations can be used to shed light on origin and effect of relaxation in the carotenoid-porphyrin-fullerene triad. The semiclassical method is then used to calculate the anisotropy in the singlet yield of simplified models of the cryptochrome radical pair, which are compared against exact quantum mechanical results, before the results of a simulation of the full cryptochrome radical pair, including all of the hyperfine interactions and electron spin coupling, are presented. The chapter concludes with a brief discussion of the recently suggested "quantum needle" of the avian compass [93].

Chapter 6 is concerned with the dynamics of the radical pairs in the semiconducting polymer layers of oLEDs which are responsible for the electroluminescence of the devices. It contains a derivation of the relationship between the singlet yield of these radical pairs and the magnetic field dependence of the electroluminescence and conductance observed experimentally in oLEDs. It then describes how the strength of the hyperfine fields of each radical may be obtained experimentally, and how singlet-triplet dephasing affects the response of electroluminescence to the application of a magnetic field. It concludes by comparing Schulten-Wolynes simulations of the magnetoelectroluminescence to the experimental results, and outlining the implications of these results.

Finally, Chap. 7 summarises the main conclusions of this thesis, and suggests two areas to which the theories presented here could be applied in the future.

References

1. Dirac, P. A. M. (1928). The quantum theory of the electron. *Proceedings of the Royal Society A: Mathematical, Physical and Engineering Sciences, 117*, 610–624.
2. Dirac, P. A. M. (1930). *The Principles of Quantum Mechanics*. Clarendon Press.
3. Slichter, C. P. (1963). *Principles of Magnetic Resonance*. Berlin: Springer.
4. Carrington, A. & McLachlan, A. D. (1967). *Introduction to Magnetic Resonance with Applications to Chemistry and Chemical Physics*. Harper & Row.
5. Atkins, P. W., & Friedman, R. S. (2011). *Molecular Quantum Mechanics*. Oxford: Oxford University Press.
6. Van Vleck, J. H. (1932). *The Theory of Electric and Magnetic Susceptibilities*. Clarendon Press.
7. Pauli, W. (1940). The connection between spin and statistics. *Physical Review, 58*, 716–722.
8. Weller, A., Staerk, H., & Treichel, R. (1984). Magnetic-field effects on geminate radical-pair recombination. *Faraday Discussions of the Chemical Society, 78*, 271.
9. Steiner, U. E., & Ulrich, T. (1989). Magnetic field effects in chemical kinetics and related phenomena. *Chemical Reviews, 89*, 51–147.

10. Fuller, G. H. (1976). Nuclear spins and moments. *Journal of Physical and Chemical Reference Data*, *5*, 835–1092.
11. Rosman, K. J. R., & Taylor, P. D. P. (1998). Isotopic compositions of the elements 1997. *Pure and Applied Chemistry*, *70*, 217–235.
12. Brocklehurst, B. (1969). Formation of excited states by recombining organic ions. *Nature*, *221*, 921–923.
13. McLauchlan, K. A., & Steiner, U. E. (1991). The spin-correlated radical pair as a reaction intermediate. *Molecular Physics*, *73*, 241–263.
14. Hore, P. J., & Mouritsen, H. (2016). The radical-pair mechanism of magnetoreception. *Annual Review of Biophysics*, *45*, 299.
15. Rodgers, C. T., & Hore, P. J. (2009). Chemical magnetoreception in birds: the radical pair mechanism. *Proceedings of the National Academy of Sciences of the United States of America*, *106*, 353–360.
16. Closs, G. L. (1969). A mechanism explaining nuclear spin polarizations in radical combination reactions. *Journal of the American Chemical Society*, *91*, 4552–4554.
17. Kaptein, R., & Oosterhoff, J. (1969). Chemically induced dynamic nuclear polarization II. *Chemical Physics Letters*, *4*, 195–197.
18. Brocklehurst, B., & McLauchlan, K. A. (1996). Free radical mechanism for the effects of environmental electromagnetic fields on biological systems. *International Journal of Radiation Biology*, *69*, 3–24.
19. Till, U., & Hore, P. J. (1997). Radical pair kinetics in a magnetic field. *Molecular Physics*, *90*, 289–296.
20. Schulten, K., & Wolynes, P. G. (1978). Semi-classical description of electron-spin motion in radicals including effect of electron hopping. *Journal of Chemical Physics*, *68*, 3292–3297.
21. Weller, A., Nolting, F., & Staerk, H. (1983). A quantitative interpretation of the magnetic field effect on hyperfine-coupling-induced triplet fromation from radical ion pairs. *Chemical Physics Letters*, *96*, 24–27.
22. Klein, J., & Voltz, R. (1976). Time-resolved optical detection of coherent spin motion for organic-radical-ion pairs in solution. *Physical Review Letters*, *36*, 1214–1217.
23. Werner, H. J., Staerk, H., & Weller, A. (1978). Solvent, isotope, and magnetic field effects in the geminate recombination of radical ion pairs. *Journal of Chemical Physics*, *68*, 2419.
24. Brocklehurst, B. (1976). Magnetic field effect on the pulse shape of scintillations due to geminate recombination of ion pairs. *Chemical Physics Letters*, *44*, 245–248.
25. Timmel, C. R., Till, U., Brocklehurst, B., Mclauchlan, K. A., & Hore, P. J. (1998). Effects of weak magnetic fields on free radical recombination reactions. *Molecular Physics*, *95*, 71–89.
26. Werner, H. J., Schulten, Z., & Schulten, K. (1977). Theory of the magnetic field modulated geminate recombination of radical ion pairs in polar solvents: application to the pyrene-N. N-dimethylaniline system. *Journal of Chemical Physics*, *67*, 646.
27. Fischer, H. (1983). The effect of a magnetic field on the product yield of a geminate radical-pair reaction in homogeneous solution. *Chemical Physics Letters*, *100*, 255–258.
28. Hamilton, C. A., Hewitt, J. P., McLauchlan, K. A., & Steiner, U. E. (1988). High resolution studies of the effects of magnetic fields on chemical reactions. *Molecular Physics*, *65*, 423–438.
29. Batchelor, S. N., Kay, C. W. M., McLauchlan, K. A., & Shkrob, I. A. (1993). Time-resolved and modulation methods in the study of the effects of magnetic fields on the yields of free-radical reactions. *Journal of Chemical Physics*, *97*, 13250–13258.
30. Brocklehurst, B. (1976). Spin correlation in the geminate reconibination of radical ions in hydrocarbons. *Journal of the Chemical Society, Faraday Transactions*, *2*(72), 1869.
31. Till, U., Timmel, C. R., Brocklehurst, B., & Hore, P. J. (1998). The inuence of very small magnetic fields on radical recombination reactions in the limit of slow recombination. *Chemical Physics Letters*, *298*, 7–14.
32. Norris, J. R., Lin, C. P., & Budil, D. E. (1987). Magnetic resonance of ultrafast chemical reactions. *Journal of the Chemical Society, Faraday Transactions*, *1*(83), 13–27.
33. Hamilton, C., McLauchlan, K., & Peterson, K. (1989). J-resonances in MARY and RYDMR spectra from freely diffusing radical ion pairs. *Chemical Physics Letters*, *162*, 145–151.

34. Weiss, E. A., Ratner, M. A., & Wasielewski, M. R. (2003). Direct measurement of singlet - triplet splitting within rodlike photogenerated radical ion pairs using magnetic field effects: estimation of the electronic coupling for charge recombination. *Journal of Physical Chemistry*, *107*, 3639–3647.
35. Weiss, E. A., Tauber, M. J., Ratner, M. A., & Wasielewski, M. R. (2005). Electron spin dynamics as a probe of molecular dynamics: temperature-dependent magnetic field effects on charge recombination within a covalent radical ion pair. *Journal of the American Chemical Society*, *127*, 6052–6061.
36. Staerk, H., Treichel, R., & Weller, A. (1983). Life uncertainty broadening in photoinduced electron transfer. *Chemical Physics Letters*, *96*, 28–30.
37. Anderson, P. W. (1959). New approach to the theory of superexchange interactions. *Physical Review*, *115*, 2–13.
38. McConnell, H. M. (1961). Intramolecular charge transfer in aromatic free radicals. *Journal of Chemical Physics*, *35*, 508–515.
39. Wasielewski, M. R. (2006). Energy, charge, and spin transport in molecules and self-assembled nanostructures inspired by photosynthesis. *Journal of Organic Chemistry*, *71*, 5051–5066.
40. Fromme, P., et al. (2002). Functional implications on the mechanism of the function of photosystem II including water oxidation based on the structure of photosystem II. *Philosophical Transactions of the Royal Society of London. Series B, Biological Sciences*, *357*, 1337–44.
41. Parson, W. W. (2003). Electron donors and acceptors in the initial steps of photosynthesis in purple bacteria: a personal account. *Photosynthesis Research*, *76*, 81–92.
42. Weiss, E. A., et al. (2005). Conformationally gated switching between superexchange and hopping within oligo-p-phenylene-based molecular wires. *Journal of the American Chemical Society*, *127*, 11842–11850.
43. Goldsmith, R. H., et al. (2005). Wire-like charge transport at near constant bridge energy through uorene oligomers. *Proceedings of the National Academy of Sciences of the United States of America*, *102*, 3540–5.
44. Jortner, J., & Ratner, M. A. (1997). *Molecular Electronics*. Oxford: Blackwell Science.
45. Nitzan, A. (2001). Electron transmission through molecules and molecular interfaces. *Annual Review of Physical Chemistry*, *52*, 681–750.
46. Moser, C. C., Keske, J. M., Warncke, K., Farid, R. S., & Dutton, P. L. (1992). Nature of biological electron transfer. *Nature*, *355*, 796–802.
47. Farid, R. S., Moser, C. C., & Dutton, P. L. (1993). Electron transfer in proteins. *Current Opinion in Structural Biology*, *3*, 225–233.
48. Murphy, C. J., et al. (1994). Fast photoinduced electron transfer through DNA intercalation. *Proceedings of the National Academy of Sciences of the United States of America*, *91*, 5315–9.
49. Davis, W. B., Svec, W. A., Ratner, M. A., & Wasielewski, M. R. (1998). Molecular-wire behaviour in p-phenylenevinylene oligomers. *Nature*, *396*, 60–63.
50. Weiss, E. A. (2004). Making a molecular wire: charge and spin transport through. *Journal of The American Chemical Society*, *126*, 5577–5584.
51. Tauber, M. J., Kelley, R. F., Giaimo, J. M., Rybtchinski, B., & Wasielewski, M. R. (2006). Electron hopping in pi-stacked covalent and self-assembled perylene diimides observed by ENDOR spectroscopy. *Journal of the American Chemical Society*, *128*, 1782–1783.
52. Scott, A. M., & Wasielewski, M. R. (2011). Temperature dependence of spin-selective charge transfer pathways in donor-bridge-acceptor molecules with oligomeric uorenone and p - phenylethynylene bridges. *Journal of the American Chemical Society*, *133*, 3005–3013.
53. Von Middendorf, A. (1859). Die Isepiptesen Rufflands. Mem. Acad. Sci. St Petersbourg VI. *Ser. Tome*, *8*, 1–143.
54. Viguier, C. Le. (1882). Sens de L'orientation et ses Organes chez les Animaux et chez L'homme. *Revue Philosophique de la France et de l' Etranger*, *14*, 1–36.
55. Wiltschko, W. (1968). Über den Einflu statischer Magnetfelder auf die Zugorientierung der Rotkehlchen (Erithacus rubecula). *Zeitschrift für Tierpsychologie*, *25*, 537–558.
56. Walker, M. M., et al. (1997). Structure and function of the vertebrate magnetic sense. *Nature*, *390*, 371–376.

57. Lohmann, K. J., Lohmann, C. M. F., & Putman, N. F. (2007). Magnetic maps in animals: nature's GPS. *Journal of Experimental Biology*, *210*, 3697–3705.
58. Mouritsen, H. (2013). *Neurosciences - From Molecule to Behavior: A University Textbook* (pp. 427–443). Berlin: Springer.
59. Schulten, K., & Windemuth, A. (1986). *Biophysical Effects of Steady Magnetic Fields* (pp. 99–106). Berlin: Springer.
60. Ritz, T., Adem, S., & Schulten, K. (2000). A model for photoreceptor-based magnetoreception in birds. *Biophysical Journal*, *78*, 707–718.
61. Wiltschko, W., & Wiltschko, R. (1972). Magnetic compass of European robins. *Science*, *176*, 62–64.
62. Wiltschko, W., & Wiltschko, R. (1996). Magnetic orientation in birds. *Journal of Experimental Biology*, *199*, 29–38.
63. Akesson, S., Morin, J., Muheim, R., & Ottosson, U. (2001). Avian orientation at steep angles of inclination: experiments with migratory white-crowned sparrows at the magnetic North Pole. *Proceedings of the Royal Society B: Biological Sciences*, *268*, 1907–1913.
64. Wiltschko, R., Ritz, T., Stapput, K., Thalau, P., & Wiltschko, W. (2005). Two different types of light-dependent responses to magnetic fields in birds. *Current Biology*, *15*, 1518–1523.
65. Wiltschko, R., et al. (2015). Magnetoreception in birds: the effect of radio-frequency fields. *Journal of the Royal Society Interface*, *12*, 20141103.
66. Maeda, K., et al. (2008). Chemical compass model of avian magnetoreception. *Nature*, *453*, 387–390.
67. Maeda, K., et al. (2012). Magnetically sensitive light-induced reactions in cryptochrome are consistent with its proposed role as a magnetoreceptor. *Proceedings of the National Academy of Sciences of the United States of America*, *109*, 4774–4779.
68. Dodson, C. A., Hore, P. J., & Wallace, M. I. (2013). A radical sense of direction: signaling and mechanism in cryptochrome magnetoreception. *Trends in Biochemical Sciences*, *38*, 435–446.
69. Lin, C., & Todo, T. (2005). The cryptochromes. *Genome Biology*, *6*, 220.
70. Nießner, C. (2011). Avian ultraviolet/violet cones identified as probable magnetoreceptors. *PLOS ONE*, *6*, e20091.
71. Solov'yov, I. A., Chandler, D. E., & Schulten, K. (2007). Magnetic field effects in arabidopsis thaliana cryptochrome-1. *Biophysical Journal*, *92*, 2711–2726.
72. Lee, A. A., et al. (2014). Alternative radical pairs for cryptochrome-based magnetoreception. *Journal of the Royal Society Interface*, *11*, 20131063.
73. Worster, S., Kattnig, D. R., & Hore, P. J. (2016). Spin relaxation of radicals in cryptochrome and its role in avian magnetoreception. *Journal of Chemical Physics*, *145*, 035104.
74. Kattnig, D. R., Solov'yov, I. A., & Hore, P. J. (2016). Electron spin relaxation in cryptochrome-based magnetoreception. *Physical Chemistry Chemical Physics*, *18*, 12443–12456.
75. Kattnig, D. R., Sowa, J. K., Solov'yov, I. A., & Hore, P. J. (2016). Electron spin relaxation can enhance the performance of a cryptochrome-based magnetic compass sensor. *New Journal of Physics*, *18*, 063007.
76. Tang, C. W., & Vanslyke, S. A. (1987). Organic electroluminescent diodes. *Applied Physics Letters*, *51*, 913–915.
77. Kido, J. (1999). Organic displays. *Physics World*, *12*, 27–30.
78. OLED displays and organic photovoltaics. (2009). *Nature Photonics*, *3*, 457.
79. Forrest, S. R. (2004). The path to ubiquitous and low-cost organic electronic appliances on plastic. *Nature*, *428*, 911–918.
80. Lupton, J. M., McCamey, D. R., & Boehme, C. (2010). Coherent spin manipulation in molecular semiconductors: getting a handle on organic spintronics. *ChemPhysChem*, *11*, 3040–3058.
81. Reineke, S. (2015). Complementary LED technologies. *Nature Materials*, *14*, 459–462.
82. Frankevich, E. L., et al. (1992). Polaron-pair generation in poly(phenylene vinylenes). *Physical Review B*, *46*, 9320–9324.
83. Hu, B., & Wu, Y. (2007). Tuning magnetoresistance between positive and negative values in organic semiconductors. *Nature Materials*, *6*, 985–91.

84. Bobbert, P. A., Nguyen, T. D., Van Oost, F. W. A., Koopmans, B., & Wohlgenannt, M. (2007). Bipolaron mechanism for organic magnetoresistance. *Physical Review Letters*, *99*, 216801.
85. Lupton, J. M., & Boehme, C. (2008). Magnetoresistance in organic semiconductors. *Nature Materials*, *7*, 598.
86. Nguyen, T. D., Gautam, B. R., Ehrenfreund, E., & Vardeny, Z. V. (2010). Magnetoconductance response in unipolar and bipolar organic diodes at ultrasmall fields. *Physical Review Letters*, *105*, 166804.
87. Ehrenfreund, E., & Vardeny, Z. V. (2012). Effects of magnetic field on conductance and electroluminescence in organic devices. *Israel Journal of Chemistry*, *52*, 552–562.
88. Cox, M., et al. (2014). Spectroscopic evidence for trap-dominated magnetic field effects in organic semiconductors. *Physical Review B*, *90*, 155205.
89. Kersten, S. P., Schellekens, A. J., Koopmans, B., & Bobbert, P. A. (2011). Magnetic-field dependence of the electroluminescence of organic light-emitting diodes: a competition between exciton formation and spin mixing. *Physical Review Letters*, *106*, 197402.
90. Nguyen, T. D., et al. (2010). Isotope effect in spin response of pi-conjugated polymer films and devices. *Nature Materials*, *9*, 345–352.
91. McCamey, D. R., Lee, S. Y., Paik, S. Y., Lupton, J. M., & Boehme, C. (2010). Spin-dependent dynamics of polaron pairs in organic semiconductors. *Physical Review B*, *82*, 125206.
92. Wang, F., Yang, C. G., Ehrenfreund, E., & Vardeny, Z. V. (2010). Spin dependent reactions of polaron pairs in PPV-based organic diodes. *Synthetic Metals*, *160*, 297–302.
93. Hiscock, H. G., et al. (2016). The quantum needle of the avian magnetic compass. *Proceedings of the National Academy of Sciences of the United States of America*, *113*, 201600341.

Chapter 2
Quantum Mechanics

In this chapter, we shall outline the quantum mechanical description of the spin dynamics of a radical pair reaction. To begin, we will sketch the origin of the Hamiltonian which governs the evolution of the electron and nuclear spins in a radical pair. We will then describe the Haberkorn operator used to account for the recombination of radical pairs, and introduce the ensemble dynamics of the radical pair, which provide a connection to experimental measurements. Using this machinery, we shall demonstrate an efficient method of calculating these ensemble averages and discuss its limitations, which motivate the development of approximate semiclassical theories of spin dynamics in Chap. 3. We will then introduce spin correlation tensors, which may be used in the special case where there is no coupling between electron spins to further reduce the computational time required to simulate radical pair reactions. Finally, we shall discuss relaxation effects, noting that the difficulty of including them in fully quantum mechanical simulations provides additional motivation to find semiclassical models which can account for these effects.[1]

2.1 The Hamiltonian

The Hamiltonian which describes the evolution of the electron and nuclear spins in the radical pairs considered in this thesis is [1, 2]

$$\begin{aligned} \hat{H} &= \hat{H}_1 + \hat{H}_2 + \hat{\mathbf{S}}_1 \cdot \mathbf{D}' \cdot \hat{\mathbf{S}}_2, \\ \hat{H}_i &= \boldsymbol{\omega}_i \cdot \hat{\mathbf{S}}_i + \sum_{k=1}^{N_i} \hat{\mathbf{I}}_{ik} \cdot \mathbf{A}'_{ik} \cdot \hat{\mathbf{S}}_i . \end{aligned} \tag{2.1}$$

[1]Parts of this chapter have been reproduced with permission from Lewis, A. M., Fay, T. P. & Manolopoulos, D. E. *Journal of Chemical Physics* **145**, 244101 (2016), https://aip.scitation.org/doi/10.1063/1.4972277

A. Lewis, *Spin Dynamics in Radical Pairs*, Springer Theses,
https://doi.org/10.1007/978-3-030-00686-0_2

In these equations, $\omega_i = -\gamma_i \mathbf{B}$, where γ_i is the gyromagnetic ratio of the electron in radical i and $\mathbf{B}$ is the applied magnetic field. $\mathbf{A}'_{ik}$ is the hyperfine coupling tensor between the kth nuclear spin and the electron spin in radical i, and $\mathbf{D}'$ is the coupling tensor between the two electron spins, which accounts for both the dipolar coupling $\mathbf{D}$ and the exchange coupling J. $\hat{\mathbf{I}}_{ik}$ and $\hat{\mathbf{S}}_i$ are the nuclear and electron spin operators respectively, and N_i is the number of nuclear spins in radical i.

In the uncoupled spin basis, this Hamiltonian is sparse; that is, most of its elements are zero. This is a result of the structure of the Hamiltonian: each term connects a given state only to states in which the projection onto the z axis of at most two spins has changed. Therefore, each state is coupled to just a small fraction of the full basis of states. There are a total of $4Z$ diagonal elements, which correspond to the terms in the Hamiltonian which leave the spin states of all of the electrons and nuclei in the radical pair unchanged. Here $Z = \Pi_{i,k}(2I_{ik} + 1)$ is the total number of nuclear spin states of the radical pair. The terms in the Hamiltonian which change the spin state of one of the electrons in the radical pair produce $8Z$ non-zero elements, with a further $4Z$ arising from terms which change the state of both electron spins simultaneously. There are a total of

$$\frac{4Z}{2I_{ik} + 1}(2I_{ik} + 1 - 1) \times 2 = 8Z \frac{2I_{ik}}{2I_{ik} + 1} \tag{2.2}$$

non-zero elements which involve changing the spin state of the kth nucleus on radical i only, and another $8Z(2I_{ik}/(2I_{ik} + 1))$ which correspond to changing the spin state of both the kth nucleus and the electron spin on radical i. Therefore, the total number of non-zero matrix elements of the Hamiltonian in the uncoupled spin basis is

$$16Z + 16Z \left(\sum_{i=1}^{2} \sum_{k=1}^{N_i} \frac{2I_{ik}}{2I_{ik} + 1} \right). \tag{2.3}$$

If $I_{ik} = 1/2$ for every nuclear spin, this expression simplifies to

$$16Z + 8Z \log_2 Z. \tag{2.4}$$

This feature of the Hamiltonian allows the efficient implementation of matrix-vector multiplications which require only $\mathcal{O}(Z \log Z)$ operations, rather than $\mathcal{O}(Z^2)$. This will be crucial when calculating the spin dynamics of molecular wires in Chap. 4.

We have neglected the contributions to the Hamiltonian from the motion of the nuclei. Translational motion can safely be ignored, since Eq. (2.1) is unaffected by a displacement. The Hamiltonian is not invariant to rotations, however – the hyperfine tensors and electron spin coupling tensor are, in general, anisotropic. If the rotational motion is rapid relative to the timescale of the spin dynamics, any anisotropic tensors will be averaged over all orientations of the radical pair, destroying them and leaving only isotropic contributions to the hyperfine and electron spin couplings, as shown in Appendix B. For each of the applications considered in this thesis, we shall assume either that the radical pairs are tumbling rapidly, or not rotating at all. The relaxation associated with the modulation of hyperfine tensors by rotational and vibrational motion is more difficult to describe, and is discussed in more detail in Sect. 2.7.1.

2.1.1 Exchange Coupling

The coupling between the electron spins is described in the Hamiltonian of Eq. (2.1) by the tensor $\mathbf{D}'$. This can be split into an isotropic component, given by the exchange coupling constant, J, and a traceless anisotropic component arising from dipolar coupling, $\mathbf{D}$:

$$\mathbf{D}' = \mathbf{D} + 2J\mathbf{E}, \tag{2.5}$$

where $\mathbf{E}$ is a 3×3 unit matrix. The exchange coupling is a consequence of the Pauli principle [3], which states that particles with half integer spin, such as electrons, follow Fermi-Dirac statistics. We shall now outline the origin of the exchange interaction, following Van Vleck in Ref. [4].

Firstly, consider a system of two non-interacting electrons which experience the same potential, $\hat{V}(\mathbf{r})$. The Schrödinger equation for this system is

$$-\frac{\hbar^2}{2m}\left(\nabla_1^2 + \nabla_2^2\right)\Psi + \left(\hat{V}(\mathbf{r}_1) + \hat{V}(\mathbf{r}_2)\right)\Psi = E_0\Psi, \tag{2.6}$$

where ∇_i^2 is the Laplacian, $\mathbf{r}_i$ is the position of particle i, and E_0 is the energy of the system. Neglecting the spin of the electrons for now, Eq. (2.6) has the degenerate solutions

$$\begin{aligned} \Psi_I &= \psi_k(\mathbf{r}_1)\psi_m(\mathbf{r}_2), \\ \Psi_{II} &= \psi_k(\mathbf{r}_2)\psi_m(\mathbf{r}_1), \\ E_0 &= E_k + E_m, \end{aligned} \tag{2.7}$$

where ψ_k and ψ_m are solutions of the Schrödinger equation for a single electron in the potential $\hat{V}(\mathbf{r})$. We will assume for simplicity that these functions are real and orthonormal. Any linear combination of Ψ_I and Ψ_{II} also satisfies Eq. (2.6).

We then introduce a perturbation to this system: a potential $\hat{V}_{12}$ between the two electrons. This breaks the degeneracy of the system, with the energy levels now given to first order by

$$\begin{vmatrix} E_0 + K_{12} - E & J_{12} \\ J_{12} & E_0 + K_{12} - E \end{vmatrix} = 0, \tag{2.8}$$

where

$$\begin{aligned} K_{12} &= \langle \Psi_I | \hat{V}_{12} | \Psi_I \rangle = \langle \Psi_{II} | \hat{V}_{12} | \Psi_{II} \rangle, \\ J_{12} &= \langle \Psi_I | \hat{V}_{12} | \Psi_{II} \rangle = \langle \Psi_{II} | \hat{V}_{12} | \Psi_I \rangle. \end{aligned} \tag{2.9}$$

Note that both J_{12} and K_{12} are positive. The solutions to Eq. (2.8) are

$$\begin{aligned} E_{+} &= E_0 + K_{12} + J_{12} \\ E_{-} &= E_0 + K_{12} - J_{12}, \end{aligned} \tag{2.10}$$

which correspond to the symmetric and antisymmetric combinations of ψ_I and ψ_{II} respectively,

$$\begin{aligned} \Psi_{+} &= \frac{1}{\sqrt{2}}\left(\Psi_I + \Psi_{II}\right), \\ \Psi_{-} &= \frac{1}{\sqrt{2}}\left(\Psi_I - \Psi_{II}\right). \end{aligned} \tag{2.11}$$

We must now introduce the spins of the electrons. Assuming that there is no spin-orbit coupling, the total wavefunction can simply be written as a product of a spatial wavefunction, such as those in Eq. (2.11), and a spin wavefunction, such as those in Eq. (1.3). In order to satisfy the Pauli principle, the total wavefunction must be antisymmetric with respect to exchange of the two electrons, so one and only one of the spin and spatial wavefunctions must be antisymmetric. As is clear from Eq. (1.3), the spin wavefunctions are antisymmetric when the total spin quantum number $S = 0$ and symmetric when $S = 1$. These must therefore be paired with the the spatial wavefunctions Ψ_{+} and Ψ_{-}, respectively, which are eigenfunctions of $\hat{V}_{12}$ with eigenvalues $K_{12} + J_{12}$ and $K_{12} - J_{12}$; the singlet state ($S = 0$) is $2J_{12}$ higher in energy than the triplet state ($S = 1$).

The eigenvalues of the square of the electron spin vector operators $\hat{\mathbf{S}}_1^2$ and $\hat{\mathbf{S}}_2^2$ are $3/4\hbar^2$, and the operator $\hat{\mathbf{S}}^2 = \left(\hat{\mathbf{S}}_1 + \hat{\mathbf{S}}_2\right)^2$ has eigenvalues of 0 and $2\hbar^2$. Therefore, the operator

$$\hat{\mathbf{S}}_1 \cdot \hat{\mathbf{S}}_2 = \frac{1}{2}\left(\hat{\mathbf{S}}^2 - \hat{\mathbf{S}}_1^2 - \hat{\mathbf{S}}_1^2\right) \tag{2.12}$$

has eigenvalues $-3/4\hbar^2$ and $1/4\hbar^2$, corresponding to when $S = 0$ and $S = 1$ respectively. Combining this with the result above, and defining $J = -J_{12}/\hbar^2$, we can say that in the basis of singlet and triplet states the operator $\hat{V}_{12} - 2J\hat{\mathbf{S}}_1 \cdot \hat{\mathbf{S}}_2 - K_{12} + 1/2 J_{12}$ is diagonal with eigenvalues equal to zero – it is a zero matrix. Since it must therefore be invariant under a similarity transform, it is generally true that

$$2J\hat{\mathbf{S}}_1 \cdot \hat{\mathbf{S}}_2 = \hat{V}_{12} - K_{12} + \frac{1}{2}J_{12}. \tag{2.13}$$

What we have shown in Eq. (2.13) is that the electrons have an effective isotropic coupling between their spins, which in fact is a consequence of the orbital parts of the wavefunction [5]. The triplet ($S = 1$) spin states must be combined with the antisymmetric spatial wavefunction, which vanishes when $\mathbf{r}_1 = \mathbf{r}_2$. This decrease in the probability density when the two electrons are close together in space is referred to as a Fermi hole; a Fermi heap is observed in the symmetric spatial wavefunction as the probability density increases when the electrons are close together [6]. The existence of a Fermi hole reduces the average electron-electron repulsion energy, leading to the difference in energy between the singlet and triplet states. Despite the orbital origin of the exchange coupling constant J, it is extremely helpful to think of it as an effective spin coupling; and it has a profound effect on the spin dynamics in a radical pair.

In addition to this "direct" exchange mechanism, there are also contributions to J from what is called the superexchange mechanism. This is a perturbative correction to the exchange coupling, which accounts for the exchange interactions in excited spin states of a molecule, and is mediated by the electrons in orbitals between the two spins [7, 8]. The magnitude of the superexchange coupling depends on the overlap between the orbitals occupied by two coupled spins and the bridging orbitals. Both the superexchange and direct exchange interactions are expected to have an exponential dependence on the separation between the two electrons [1, 9],

$$J(r) = J_0 e^{-(r/r_J)}, \tag{2.14}$$

where J_0 and r_J are parameters which characterise the magnitude and decay of the interaction. However, while the contribution to J from direct exchange is always negative, contributions from superexchange may be positive or negative, depending on the relative energies of the excited states involved [8]. As a result, if the superexchange mechanism outweighs the direct exchange mechanism the overall coupling constant may be positive, resulting in the singlet state being lower in energy than the triplet state.

2.1.2 Dipolar Coupling

The anisotropic component $\mathbf{D}$ of the electron spin coupling tensor $\mathbf{D}'$ in Eq. (2.5) arises from the dipolar coupling between the two electron spins. An electron with spin $\mathbf{S}_i$ has an associated magnetic dipole $\mathbf{m}_i$, related by [10]

$$\mathbf{m}_i = \gamma_i \mathbf{S}_i, \tag{2.15}$$

where γ_i is the gyromagnetic ratio of the electron. The magnetic field generated by that magnetic dipole at some point in space $\mathbf{r}$, assuming it to be a point dipole, is given by [11]

$$\mathbf{B}_i = -\frac{\mu_0}{4\pi r^3}\left(\mathbf{m}_i - \frac{3\mathbf{r}(\mathbf{r}\cdot\mathbf{m}_i)}{r^2}\right), \tag{2.16}$$

where μ_0 is the vacuum permeability, and $r = |\mathbf{r}|$. The Hamiltonian which describes the interaction of this field with another magnetic dipole is

$$\hat{H}_D = -\hat{\mathbf{B}}_1 \cdot \hat{\mathbf{m}}_2$$
$$= \frac{\gamma_1 \gamma_2 \mu_0}{4\pi r^3} \left(\hat{\mathbf{S}}_1 \cdot \hat{\mathbf{S}}_2 - \frac{3(\hat{\mathbf{S}}_1 \cdot \mathbf{r})(\mathbf{r} \cdot \hat{\mathbf{S}}_2)}{r^2} \right) \tag{2.17}$$

provided that the dipoles are far apart [12]. We may rewrite this expression as

$$\hat{H}_D = \hat{\mathbf{S}}_1 \cdot \mathbf{D} \cdot \hat{\mathbf{S}}_2, \tag{2.18}$$

where $\mathbf{D}$ is a 3×3 matrix given by

$$\mathbf{D} = \frac{\gamma_1 \gamma_2 \mu_0}{4\pi r^3} \left(\mathbf{E} - 3\frac{\mathbf{r} \otimes \mathbf{r}}{r^2} \right)$$
$$= D \left(\mathbf{E} - 3\frac{\mathbf{r} \otimes \mathbf{r}}{r^2} \right), \tag{2.19}$$

$\mathbf{E}$ is the unit matrix, and $\otimes$ indicates an outer product.

It is important to note that the dipolar coupling tensor vanishes if the radical pair is tumbling in solution; this is shown explicitly in Appendix B. The dipolar coupling is sometimes referred to as the zero-field splitting, since it lifts the degeneracy of the triplet states of a radical pair even in the absence of an external magnetic field. However, it is not the only possible origin of a zero-field splitting; spin orbit coupling, if present, has a similar symmetry-breaking effect [13].

2.1.3 The Zeeman Interaction

The first term of $\hat{H}_i$ in Eq. (2.1) is the Zeeman interaction between an electron spin and an external magnetic field $\mathbf{B}$ [5],

$$\hat{H}_Z = \boldsymbol{\omega} \cdot \hat{\mathbf{S}} = -\gamma \mathbf{B} \cdot \hat{\mathbf{S}} = -\mathbf{B} \cdot \hat{\mathbf{m}}. \tag{2.20}$$

The gyromagnetic ratio of the electron, γ, is defined as $-g\mu_B/\hbar$, where μ_B is the Bohr magneton and g is the electron spin g-factor. In this work, we assume that the g-factor is isotropic and equal to that of the free electron, g_e, and hence that the gyromagnetic ratio of an electron is γ_e. The first assumption is easily justified if the system under investigation is tumbling in solution, since any anisotropic component of the g-tensor will average to zero, and so not influence the spin dynamics of the radical pair [13].

Even when molecular tumbling is slow, the g-tensor will only be anisotropic if there is significant coupling between the electron's spin and orbital angular momentum [12]. This is generally not observed in the type of planar, conjugated radicals we consider in this thesis [14]. This also justifies our second assumption, since the isotropic g-factors of electrons do not vary significantly from the free electron value without strong spin-orbit coupling [12]. To second order in perturbation theory, the isotropic g-factor is given by [13]

$$g = g_e - \frac{2\lambda}{3} \sum_{n>0} \frac{\langle 0|\hat{\mathbf{L}}|n\rangle \cdot \langle n|\hat{\mathbf{L}}|0\rangle}{E_n - E_0}, \tag{2.21}$$

where λ is the spin-orbit coupling constant, $\hat{\mathbf{L}}$ is the electron orbital angular momentum vector operator, $|n\rangle$ is an eigenstate of the spatial Hamiltonian, and E_n is the energy of that state. It is clear from this expression that $g \approx g_e$ unless λ is large relative to the difference between the energies of the spatial eigenstates, $E_n - E_0$.

The Hamiltonian in Eq. (2.1) neglects the Zeeman interactions between the magnetic field and the nuclear spins. This approximation is based on the fact that these interactions are at least three orders of magnitude smaller than the electronic Zeeman interactions, since the gyromagnetic ratio of a particle is inversely proportional to its mass [13].

2.1.4 Hyperfine Interactions

The final term of $\hat{H}_i$ in Eq. (2.1) describes the hyperfine interactions between the electron spin and nuclear spins in radical i. These couplings have an isotropic component a, arising from the Fermi contact interaction, and a traceless anisotropic part $\mathbf{A}$ from the dipolar coupling between the electron and nuclear spins [13]. The total hyperfine tensor is defined as

$$\mathbf{A}' = \mathbf{A} + a\mathbf{E}, \tag{2.22}$$

where $\mathbf{E}$ is a 3×3 unit matrix. The anisotropic part of the hyperfine interaction $\mathbf{A}$ is entirely analogous to the tensor which describes the dipolar coupling between two electrons, $\mathbf{D}$. We shall therefore not repeat that discussion here, other than to re-emphasise that the tensor $\mathbf{A}$ averages to zero for radicals tumbling in solution. Instead, we will focus on the origin of the isotropic coupling constant, a.

The Fermi contact interaction between a nuclear spin and an electron spin is a correction to point dipole approximation made in our derivation of dipolar coupling. Eq. (2.17) has a singularity at $r = 0$, and so does not accurately describe systems which have non-zero electron density at the position of the nucleus. We shall consider the simplest example for which this correction is required, an electron in the $1s$ orbital of a hydrogen atom. Firstly, we define the average of the magnetic field $\mathbf{B}$ generated by the magnetic moment of the nucleus $\mathbf{m}_N$ over the electron probability density, $|\Psi(\mathbf{r})|^2$,

$$\overline{\mathbf{B}} = \int \mathbf{B}(\mathbf{r})|\Psi(\mathbf{r})|^2 d\mathbf{r}. \tag{2.23}$$

Taking the nucleus to be a sphere of radius b, and noting that $|\Psi(\mathbf{r})|^2$ is spherically symmetric, it is clear that the contribution to the integral when $r > b$ will vanish, but a term corresponding to a uniform field $\mathbf{B}_c$ within $r < b$ will survive [12]. If $|\Psi(\mathbf{r})|^2$ is slowly varying within the volume of the nucleus, then

$$\overline{\mathbf{B}} = \mathbf{B}_c|\Psi(0)|^2\frac{4\pi}{3}b^3. \tag{2.24}$$

$\mathbf{B}_c$ can be shown to be equal to [12]

$$\mathbf{B}_c = \frac{2\mu_0\mathbf{m}_N}{4\pi b^3}, \tag{2.25}$$

which may be substituted back into Eq. (2.24), leaving

$$\overline{\mathbf{B}} = \frac{2}{3}\,\mu_0\mathbf{m}_N|\Psi(0)|^2. \tag{2.26}$$

The interaction of this averaged field with the magnetic moment of an electron $\mathbf{m}_e$ is described by the spin Hamiltonian

$$\begin{aligned}\hat{H}_F &= -\hat{\overline{\mathbf{B}}}\cdot\hat{\mathbf{m}}_e\\ &= -\frac{2}{3}\mu_0(\hat{\mathbf{m}}_N\cdot\hat{\mathbf{m}}_e)|\Psi(0)|^2\\ &= -\frac{2}{3}\mu_0\gamma_e\gamma_N\,\hat{\mathbf{I}}\cdot\hat{\mathbf{S}}|\Psi(0)|^2,\end{aligned} \tag{2.27}$$

where γ_e and γ_N are the gyromagnetic ratios of the free electron and nucleus, respectively. Note that this Hamiltonian is applied only to the spin part of the wavefunction; the average of the spatial wavefunction has already been taken, which leads to the term $|\Psi(0)|^2$ in Eq. (2.27). We may now define the isotropic hyperfine coupling constant

$$a = -\frac{2}{3}\mu_0\gamma_e\gamma_N\,|\Psi(0)|^2. \tag{2.28}$$

This expression can be easily generalised to a molecular hyperfine constant [13]:

$$a = -\frac{2}{3}\mu_0\gamma_e\gamma_N\rho^S(0). \tag{2.29}$$

Here $\rho^S(0)$ is the electron spin density at the nucleus of interest, and is defined as [13, 15]

$$\rho^S(0) = 2\langle\chi|\sum_k\delta(\mathbf{r}_k)\hat{\mathbf{S}}_{kz}|\chi\rangle \tag{2.30}$$

where $|\chi\rangle$ is a multi-electron wavefunction, $\mathbf{r}_k$ is the position of the kth electron, $\delta(\mathbf{r}_k)$ is the Dirac delta function, and the sum is over every electron. $\rho^S(0)$ may be simply interpreted as the difference between the average number of electrons at the nucleus in the spin states $|\alpha\rangle$ and $|\beta\rangle$. Note that $\rho^S(0)$, and hence the corresponding hyperfine constant, may be either positive or negative for a given nucleus.

Having described the origin of each term in the Hamiltonian which describes the spin evolution of a radical pair, we will now discuss the quantum mechanical operator used to describe their recombination reactions.

2.2 The Recombination Operator

The reactions of radical pairs are typically spin selective: the rate and products of the reaction depend on the spin state of the radical pair. As a result, the recombination of radical pairs is dynamically coupled to their Hamiltonian evolution, and cannot in general be factored out of the spin dynamics. Section 2.2.1 outlines some properties of the Haberkorn operator used throughout this thesis to account for recombination. Section 2.2.2 discusses and critiques an alternative recombination operator which has been proposed.

2.2.1 The Haberkorn Recombination Operator

The Haberkorn recombination operator is defined as [16]

$$\hat{K} = \frac{k_\text{S}}{2}\hat{P}_\text{S} + \frac{k_\text{T}}{2}\hat{P}_\text{T}, \tag{2.31}$$

where

$$\hat{P}_\text{S} = \frac{1}{4}\hat{1} - \hat{\mathbf{S}}_1 \cdot \hat{\mathbf{S}}_2, \tag{2.32}$$

$$\hat{P}_\text{T} = \frac{3}{4}\hat{1} + \hat{\mathbf{S}}_1 \cdot \hat{\mathbf{S}}_2, \tag{2.33}$$

are the projection operators onto the singlet and triplet subspaces, and k_S and k_T are the first order rate constants for recombination of the singlet and triplet states, respectively. In the absence of any other spin evolution ($\hat{H} = 0$), the density operator $\hat{\rho}$ evolves under this recombination operator according to

$$\frac{\text{d}\hat{\rho}}{\text{d}t} = -\left(\hat{K}\hat{\rho} + \hat{\rho}\hat{K}\right) = -\left\{\hat{K}, \hat{\rho}\right\}, \tag{2.34}$$

where $\{\cdots\}$ indicates an anticommutator.

To illustrate the effect of the recombination operator, we will consider a simple example: the recombination of a pure state $|\alpha_1\beta_2\rangle$. In the $|S, M_S\rangle$ basis $\{|1,0\rangle, |0,0\rangle\}$, the density matrix which describes this state is

$$\hat{\rho}(0) = \frac{1}{2}\begin{pmatrix} 1 & 1 \\ 1 & 1 \end{pmatrix}. \tag{2.35}$$

Since the recombination operator is diagonal in this basis, the solution to Eq. (2.34) is straightforward [16]:

$$\hat{\rho}(t) = \frac{1}{2}\begin{pmatrix} e^{-k_\mathrm{T}t} & e^{-1/2(k_\mathrm{S}+k_\mathrm{T})t} \\ e^{-1/2(k_\mathrm{S}+k_\mathrm{T})t} & e^{-k_\mathrm{S}t} \end{pmatrix}. \tag{2.36}$$

This predicts the correct dynamics: the population of the triplet and singlet states decay exponentially with rate constants k_S and k_T respectively. The coherence between the two states decays exponentially with the mean of their rate constants. This may be trivially extended to the full basis of electronic spin states.

2.2.2 An Alternative Recombination Operator

Despite the Haberkorn operator being used uncontroversially for over 30 years, its validity has recently been questioned [17]. In particular, it was suggested that the Haberkorn operator failed to explain the experimental data presented by Maeda et al. which demonstrated an Earth-strength magnetic field effect [18] – the same data discussed in Sect. 5.2.[2] In response to this, the alternative Jones model of recombination was proposed based on quantum measurement theory [19].

In the language of that theory, the Haberkorn operator treats chemical reactions as continuous weak measurements; that is, measurement failure indicates nothing about the system. On the other hand, the Jones operator treats reactions as probabilistic strong measurements, where failure provides definite information about the state of the system [20]. This has the effect of dephasing the system more quickly than the Haberkorn model: applying the Jones model of recombination to the density matrix in Eq. (2.35) produces the density matrix

$$\hat{\rho}(t) = \frac{1}{2}\begin{pmatrix} e^{-k_\mathrm{T}t} & e^{-(k_\mathrm{S}+k_\mathrm{T})t} \\ e^{-(k_\mathrm{S}+k_\mathrm{T})t} & e^{-k_\mathrm{S}t} \end{pmatrix}, \tag{2.37}$$

by contrast to Eq. (2.36). Here the singlet-triplet coherences decay at the sum of the singlet and triplet recombination rates, rather than their average.

[2] In fact, in Sect. 5.2 we will show that Maeda's results *can* be reproduced using the Haberkorn operator, provided that other sources of dephasing are accounted for.

This discrepancy between the dephasing rates of the two approaches is subtle, particularly since they only provide lower bounds for the dephasing rate; usually there will be additional sources of dephasing caused by the motion of the radical pair. Therefore, differentiating between the two possible operators is very difficult to do experimentally, since isolating other sources of dephasing is challenging. Nevertheless, an experimental test of the validity of the recombination operators has been performed using the carotenoid-porphyrin-fullerene triad described in Sect. 5.2 [21]. In that case, it was found that the dephasing was inconsistent with the Jones model, but in agreement with the Haberkorn operator.

In addition, it has been shown that the Haberkorn model may be derived from the quantum mechanical treatment of a simple recombination reaction in which a radical pair interacts with a harmonic bath [22]. Numerical simulations involving more complex system-bath interactions also supported the Haberkorn model of recombination. In the light of this, and the lack of experimental evidence calling it into question, we will continue to use Haberkorn's recombination operator in this work.

2.3 Observables

Radical pairs are rarely investigated in isolation experimentally; instead, measurements are usually made on an ensemble of radical pairs. In quantum mechanics, the ensemble average of an observable $\mathrm{A}(t)$ corresponding to an operator $\hat{A}$ is given by [23]

$$\mathrm{A}(t) = \mathrm{tr}\left[\hat{\rho}(t)\hat{A}\right], \tag{2.38}$$

where $\hat{\rho}(t)$ is the density operator at time t, and the trace is taken over the entire Hilbert space. In the case that the density operator represents a pure state, this is identical to the expectation value of the operator in that state; in the case of a mixed state, it is the sum of the expectation values of each state weighted by its probability within the statistical ensemble. We shall now consider how to calculate these ensemble averages. Please note that for the remainder of this chapter we will work in a unit system where $\hbar = 1$, for convenience.

The unitary evolution of the density operator is described by the von Neumann equation [24]

$$\frac{\mathrm{d}\hat{\rho}}{\mathrm{d}t} = -i\left(\hat{H}\hat{\rho} - \hat{\rho}\hat{H}\right) = -i\left[\hat{H}, \hat{\rho}\right], \tag{2.39}$$

where $[\cdots]$ indicates a commutator. We must add to this a term describing the spin-selective recombination of radical pairs, given by Eqs. (2.31) and (2.34):

$$\frac{\mathrm{d}\hat{\rho}}{\mathrm{d}t} = -i\left[\hat{H}, \hat{\rho}\right] - \left\{\hat{K}, \hat{\rho}\right\}. \tag{2.40}$$

Solving this equation yields an expression for the density operator at time t [25]:

$$\hat{\rho}(t) = e^{-i\hat{H}t-\hat{K}t}\hat{\rho}(0)e^{+i\hat{H}t-\hat{K}t}. \tag{2.41}$$

While in theory Eq. (2.41) is all that is required to calculate the ensemble averages of observables, in practice its evaluation is not feasible for realistic radical pairs. Both $\hat{\rho}$ and the propagation operator $e^{-i\hat{H}t-\hat{K}t}$ have $(4Z)^2$ elements, and since Z scales exponentially with the number of nuclear spins, calculating and storing the matrix elements of these operators becomes prohibitive when $N > 10$.

By a little manipulation, it is possible find a less memory-intensive method to calculate ensemble averages. Defining

$$\hat{A}(t) = e^{+i\hat{H}t-\hat{K}t}\,\hat{A}\,e^{-i\hat{H}t-\hat{K}t}, \tag{2.42}$$

inserting Eq. (2.41) into (2.38), and using the invariance of a trace to cyclic permutation yields

$$\mathrm{A}(t) = \mathrm{tr}\left[\hat{\rho}(0)\hat{A}(t)\right]. \tag{2.43}$$

This trace may be evaluated in the basis

$$\{\mathcal{B}\} = \{\,|\Theta\rangle \otimes |\mathbf{M}_1\rangle \otimes |\mathbf{M}_2\rangle\}. \tag{2.44}$$

Here $|\Theta\rangle$ is the electronic spin state, and Θ can take the values S, T_+, T_0 and T_-, representing the singlet and triplet states respectively. $|\mathbf{M}_i\rangle$ is the nuclear spin state of radical i, given by

$$|\mathbf{M}_i\rangle = |M_{i1}\rangle \otimes |M_{i2}\rangle \otimes \cdots \otimes \left|M_{iN_i}\right\rangle, \tag{2.45}$$

where M_{ik} is the projection of the kth nuclear spin in radical i onto the z axis. In this basis, Eq. (2.43) becomes

$$\mathrm{A}(t) = \sum_{\Theta}\sum_{\mathbf{M}_1}\sum_{\mathbf{M}_2}\langle\Theta, \mathbf{M}_1, \mathbf{M}_2|\hat{\rho}(0)\hat{A}(t)|\Theta, \mathbf{M}_1, \mathbf{M}_2\rangle \tag{2.46}$$

where the sum over $\mathbf{M}_i$ indicates the sum over all possible nuclear spin states $|\mathbf{M}_i\rangle$.

In the case of photoexcited radical pairs formed in the singlet state, $\hat{\rho}(0) = \hat{P}_\mathrm{S}/Z$. The sum over Θ can then be evaluated immediately: only terms with $\Theta = \mathrm{S}$ survive. Then, defining

$$|\mathrm{S}, \mathbf{M}_1, \mathbf{M}_2; t\rangle = e^{-i\hat{H}t-\hat{K}t}\,|\mathrm{S}, \mathbf{M}_1, \mathbf{M}_2\rangle, \tag{2.47}$$

Equation (2.46) becomes

$$\mathrm{A}(t) = \frac{1}{Z}\sum_{\mathbf{M}_1}\sum_{\mathbf{M}_2}\langle\mathrm{S}, \mathbf{M}_1, \mathbf{M}_2; t|\hat{A}|\mathrm{S}, \mathbf{M}_1, \mathbf{M}_2; t\rangle. \tag{2.48}$$

Therefore, the ensemble average of an observable can be evaluated by propagating Z wavepackets with orthogonal initial nuclear spin states and calculating the expectation value of $\hat{A}$ in each state at time t. The propagation described in Eq. (2.47) may be performed by various different algorithms, two of which are discussed in Appendix A.

While this wavepacket propagation approach significantly reduces the memory requirements of calculating ensemble averages of observables, it is still computationally expensive. Both the number of wavepacket propagations required and the CPU time required for each propagation increase exponentially as the number of nuclear spins in the radical pair increases. In Sect.2.5, we will discuss a more efficient method which reduces the number of wavepacket propagations required. This method uses stochastic sampling of the coherent spin states of the nuclear spins, which are introduced in the following section.

2.4 Coherent Spin States

A coherent spin state of a spin with quantum number J, $|\Omega_J\rangle$, is a quantum state in which the expectation value of the angular momentum operator $\hat{\mathbf{J}}$ is given by [26]

$$\begin{aligned} \langle\Omega_J|\hat{\mathbf{J}}|\Omega_J\rangle &= J\mathbf{n}(\Omega), \\ \mathbf{n}(\Omega) &= (\sin\theta\cos\phi, \sin\theta\sin\phi, \cos\theta)^{\top}. \end{aligned} \tag{2.49}$$

They are the quantum states in which the behaviour of the expectation values of the angular momentum operator $\hat{\mathbf{J}}$ most closely resemble a classical angular momentum. $|\Omega_J\rangle$ may be obtained by rotating the quantisation axis of the $|J, J\rangle$ eigenstate of $\hat{J}_z$ to lie in the direction $\Omega = (\theta, \phi)$ [26],

$$|\Omega_J\rangle = (\cos\theta/2)^{2J} \exp\left\{\tan(\theta/2)\, e^{i\phi}\hat{J}_-\right\} |J, J\rangle, \tag{2.50}$$

where $0 \leqslant \theta \leqslant \pi$ and $0 \leqslant \phi < 2\pi$. Equivalently, $|\Omega_J\rangle$ can be expressed as a weighted sum of the eigenstates of $\hat{J}_z$ [27]:

$$|\Omega_J\rangle = \sum_{M=-J}^{J} \binom{2J}{J+M}^{1/2} (\cos\theta/2)^{J+M} (\sin\theta/2)^{J-M} e^{i(J-M)\phi} |J, M\rangle. \tag{2.51}$$

Using this definition, it can clearly be seen that the coherent spin states are not orthogonal [27]:

$$\langle\Omega'_J \mid |\Omega_J\rangle = \left(\cos\tfrac{\theta}{2}\cos\tfrac{\theta'}{2} + e^{i(\phi-\phi')}\sin\tfrac{\theta}{2}\sin\tfrac{\theta'}{2}\right)^{2J} \neq \delta(\Omega - \Omega') \tag{2.52}$$

We shall now prove that the set of states $\{|\Omega_J\rangle\}$ resolves the identity [26]:

$$\hat{1} = \frac{2J+1}{4\pi}\int d\Omega\, |\Omega_J\rangle\,\langle\Omega_J|\,. \tag{2.53}$$

Inserting the definition of a coherent spin state given in Eq. (2.51), we find [28]

$$\begin{aligned}
\hat{1} &= \frac{2J+1}{4\pi}\int_0^{2\pi} d\phi \int_0^{\pi} \sin\theta\, d\theta \Bigg[\sum_{M,M'=-J}^{J} \binom{2J}{J+M}\left(\cos\tfrac{\theta}{2}\right)^{2J+M+M'} \\
&\qquad\qquad \times \left(\sin\tfrac{\theta}{2}\right)^{2J-M-M'} e^{i(M'-M)\phi}\, |J, M'\rangle\,\langle J, M|\Bigg] \\
&= \sum_{M=-J}^{J} |J, M\rangle\,\langle J, M|\, \frac{2J+1}{2}\binom{2J}{J+M}\int_0^{\pi} \sin\theta\, d\theta\left[\left(\cos\tfrac{\theta}{2}\right)^{2J+2M}\left(\sin\tfrac{\theta}{2}\right)^{2J-2M}\right]
\end{aligned} \tag{2.54}$$

The integral on the final line of Eq. (2.54), which we shall label $\lambda_{J,M}$, can be solved using a reduction formula:

$$\begin{aligned}
\lambda_{J,M} &= 4\int_0^{\pi/2} d\tfrac{\theta}{2}\left[\left(\cos\tfrac{\theta}{2}\right)^{2(J+M)}\cos\tfrac{\theta}{2}\left(\sin\tfrac{\theta}{2}\right)^{2(J-M)+1}\right] \\
&= 4\left(\frac{J+M}{J-M+1}\right)\int_0^{\pi/2} d\tfrac{\theta}{2}\left[\left(\cos\tfrac{\theta}{2}\right)^{2(J+M)-2}\cos\tfrac{\theta}{2}\left(\sin\tfrac{\theta}{2}\right)^{2(J-M)+3}\right] \\
&= 4\left(\frac{J+M}{J-M+1}\right)\lambda_{J,M-1} \\
&= 4\left(\frac{J+M}{J-M+1}\right)\left(\frac{J+M-1}{J-M+2}\right)\cdots\left(\frac{J+1}{J}\right)\lambda_{J,0} \\
&= 4\left(\frac{(J+M)!}{J!}\right)\left(\frac{(J-M)!}{J!}\right)\lambda_{J,0}
\end{aligned} \tag{2.55}$$

The integral $\lambda_{J,0}$ may be straightforwardly evaluated

$$\begin{aligned}
\lambda_{J,0} &= \int_0^{\pi/2} d\tfrac{\theta}{2}\left(\cos\tfrac{\theta}{2}\right)^{2J+1}\left(\sin\tfrac{\theta}{2}\right)^{2J+1} \\
&= \frac{1}{2}\int_0^{\pi} d\theta\left(\tfrac{1}{2}\sin\theta\right)^{2J+1} \\
&= \frac{1}{2}\frac{(J!)^2}{(2J+1)!}
\end{aligned} \tag{2.56}$$

Finally, by combining Eqs. (2.54), (2.55), and (2.56), we can prove Eq. (2.53):

$$
\begin{aligned}
\hat{1} &= \sum_{M=-J}^{J} |J, M\rangle \langle J, M| \binom{2J}{J+M} \left(\frac{(2J+1)(J!)^2}{(2J+1)!}\right) \left(\frac{(J+M)!}{J!}\right) \left(\frac{(J-M)!}{J!}\right) \\
&= \sum_{M=-J}^{J} |J, M\rangle \langle J, M| \left(\frac{(2J)!}{(J+M)!(J-M)!}\right) \left(\frac{(J!)^2}{(2J)!}\right) \left(\frac{(J+M)!(J-M)!}{(J!)^2}\right) \\
&= \sum_{M=-J}^{J} |J, M\rangle \langle J, M|
\end{aligned}
\tag{2.57}
$$

Note that since the coherent states are not orthogonal, the set of states $\{|\Omega_J\rangle\}$ is overcomplete.

The ensemble average of an observable may be evaluated in the basis of coherent spin states. By inserting equation Eq. (2.53) into (2.48), the sums over $\mathbf{M}_1$ and $\mathbf{M}_2$ can be replaced by an $2N$ dimensional integral over the coherent spin states of each nuclear spin:

$$
\mathrm{A}(t) = \frac{1}{(4\pi)^N} \int \mathrm{d}\Omega_1 \int \mathrm{d}\Omega_2 \langle \mathrm{S}, \Omega_1, \Omega_2; t|\hat{A}|\mathrm{S}, \Omega_1, \Omega_2; t\rangle. \tag{2.58}
$$

Here

$$
|\Omega_i\rangle = |\Omega_{i1}\rangle \otimes |\Omega_{i2}\rangle \otimes \cdots \otimes \left|\Omega_{iN_i}\right\rangle, \tag{2.59}
$$

$|\Omega_{ik}\rangle$ is a coherent spin state of the kth nuclear spin of radical i, and $\int \mathrm{d}\Omega_i$ indicates the integral over all possible orientations of each of these coherent spin states. Note that the factors of $(2J+1)$ in Eq. (2.53) cancel with $1/Z$ in Eq. (2.48).

2.5 An Efficient Quantum Mechanical Method

We have shown that the ensemble average of an observable may be expressed in terms of integrals over the coherent spin state basis of each nuclear spin in a radical pair. These integrals, and hence $\mathrm{A}(t)$, may be evaluated by Monte Carlo sampling the directions of the initial nuclear spin states $|\Omega_{ik}\rangle$ from the surfaces of spheres. If the number of samples M required to converge these integrals is significantly smaller than Z, then evaluating $\mathrm{A}(t)$ using Eq. (2.58) will be far less expensive than using Eq. (2.48).

In order to investigate this, we have calculated the singlet yield of a model radical pair,

$$
\Phi_\mathrm{S} = k_\mathrm{S} \int_0^\infty \mathrm{P_S}(t)\, \mathrm{d}t. \tag{2.60}
$$

$\mathrm{P_S}(t)$ is the ensemble average of the singlet probability at time t, obtained by substituting $\hat{A} = \hat{P}_\mathrm{S}$ into Eq. (2.58). In our model system, dipolar interactions are assumed to be averaged out by tumbling in solution, and each electron is coupled to ten

Table 2.1 The hyperfine coupling constants used in the model radical pair

k	$a_{1k}/\lvert\gamma_e\rvert$ (mT)	$a_{2k}/\lvert\gamma_e\rvert$ (mT)
1	−0.999985	−0.232996
2	−0.736925	0.0388327
3	0.511210	0.661931
4	−0.0826998	−0.930856
5	0.0655341	−0.893077
6	−0.562082	0.0594001
7	−0.905911	0.342299
8	0.357729	−0.984604
9	0.358593	−0.233169
10	0.869386	−0.866316

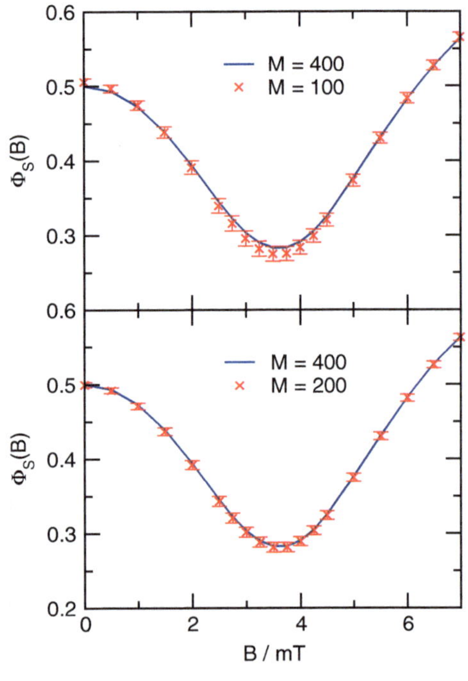

Fig. 2.1 The singlet yield of the model radical pair as a function of the applied magnetic field strength B, averaged over M initial nuclear spin states. The error bars are given by $\pm\epsilon$, the standard error in the mean of the M samples

nuclear spins with random isotropic hyperfine coupling constants uniformly distributed between $-1 < a_{ik} < 1$ mT, which are listed in Table 2.1. The exchange coupling constant is taken to be $J/|\gamma_e| = 1.75$ mT, comparable to the effective hyperfine field strength in each radical ($B_{\text{hyp},1} = 1.72$, $B_{\text{hyp},2} = 1.74$), and the rate constants are chosen as $k_\text{S}/|\gamma_e| = 0.1$ mT and $k_\text{T}/|\gamma_e| = 1.0$ mT.

Figure 2.1 shows the convergence of the singlet yield of this model radical pair with M across a wide range of magnetic field strengths. The results are converged to graphical accuracy at all field strengths with only $M = 200$ samples. Since the total number of nuclear spin states of the radical pair is $Z = 1{,}048{,}576$, this stochastic calculation is > 5000 times faster than the equivalent deterministic one would be.

The form of the MFE on the singlet yield shown in Fig. 2.1 can be understood by considering the energy levels of the spin states of the radical pair, shown schematically in Fig. 1.1. At both zero and high fields, the singlet state is separated in energy from the triplet states, which leads to slow conversion between the spin states. By contrast, when $\omega = |\boldsymbol{\omega}| = 2J$, the singlet state is degenerate with the $|\mathrm{T}_-\rangle$ state, and hyperfine mediated intersystem crossing becomes much more efficient. Since the radical pair is formed in the singlet state, any intersystem crossing will reduce the singlet yield. Therefore, a minimum is observed in the the singlet yield when $\omega = 2J$; this is the resonance effect described in Sect. 1.3.3.

This explains the observation that in the top panel of Fig. 2.1 the convergence of the Monte Carlo integration is slowest, and the error bars largest, around the minimum in the singlet yield. Away from resonance, the hyperfine interactions play only a small role in the spin dynamics, and the singlet yield does not vary significantly between Monte Carlo samples. However, when the intersystem crossing is fast the singlet yield depends strongly on the initial nuclear spin state, and more samples are required to obtain converged results.

The field dependent sensitivity of the spin dynamics to the initial nuclear spin state is illustrated by the distribution of singlet yields obtained from individual wavepackets,

$$\phi_\mathrm{S}(\boldsymbol{\Omega}_1, \boldsymbol{\Omega}_2) = k_\mathrm{S} \int_0^\infty \langle S, \boldsymbol{\Omega}_1, \boldsymbol{\Omega}_2; t|\hat{P}_\mathrm{S}|S, \boldsymbol{\Omega}_1, \boldsymbol{\Omega}_2; t\rangle \mathrm{d}t. \tag{2.61}$$

Figure 2.2 shows a histogram of 1000 evaluations of this singlet yield at three different magnetic field strengths. This highlights the increased variation in ϕ_S when intersystem crossing becomes more efficient. Nevertheless, even when $\omega = 2J$, the number of coherent spin state samples required to converge the ensemble average of the singlet yield is still far smaller than the total number of nuclear spin states, Z.

This analysis also explains why stochastic integration over coherent spin states is so efficient. Since the singlet yield is bounded between zero and one, the maximum possible standard deviation in the distribution of ϕ_S is $\sigma = 1/2$. In practice, the distribution is likely to be confined to a smaller range, as demonstrated by Fig. 2.2, and so have an even smaller standard deviation. When performing a Monte Carlo integration with random sampling, the standard deviation is related to the error in the integral, ϵ, and the number of samples, M, by [29]

$$\epsilon \propto \frac{\sigma}{\sqrt{M}}. \tag{2.62}$$

Therefore, the necessarily small standard deviation in the distribution of singlet yields means that only a small number of samples are required to converge the results. We expect this observation to be general, since the observables of interest in spin dynamics calculations are typically probabilities or yields, which must be bounded

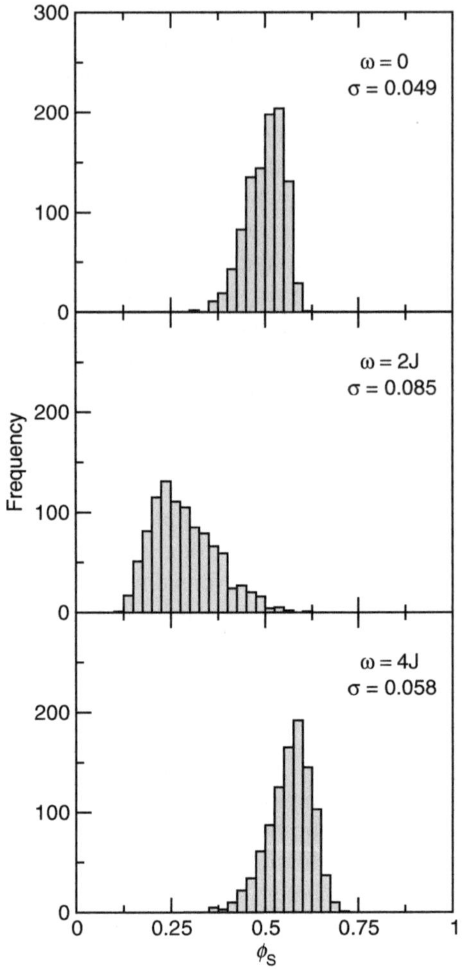

Fig. 2.2 Histograms of 1000 single wavepacket singlet yields, ϕ_S, of the model radical pair at three different applied magnetic field strengths

between zero and one. As a result, this stochastic approach should be applicable to a wide range of radical pairs. While the precise number of samples required to obtain converged results will doubtless vary from system to system, it seems likely that for radical pairs with $N > 10$ nuclear spins, M will be significantly smaller than the number of nuclear spin states.

As an alternative to the method presented here, we also considered the possibility of stochastically sampling $\mathbf{M}_1$ and $\mathbf{M}_2$ in order to evaluate the sums in Eq. (2.48). We found that convergence was much slower than that observed when sampling coherent spin states. The reason for this lies in the fact that the Hamiltonian in Eq. (2.1) commutes with the total spin projection operator, $\hat{J}_z = \hat{S}_{1z} + \hat{S}_{2z} + \sum_{i,k} \hat{I}_{ikz}$. This divides the Hilbert space into sectors of different M_J, which all contribute independently to Φ_S. Since the state $|\text{S}, \mathbf{M}_1, \mathbf{M}_2\rangle$ is an eigenstate of $\hat{J}_z$, each choice of $\mathbf{M}_1$

and $\mathbf{M}_2$ samples only a single sector of the total Hilbert space. By contrast, each coherent spin state includes some contribution from every M_J sector in the Hilbert space. If anisotropic hyperfine coupling or dipolar coupling were introduced, the Hamiltonian would no longer commute with $\hat{J}_z$, and so we expect that the convergence of results by stochastically sampling the eigenstates of $\hat{J}_z$ would be faster. Nonetheless, we would still expect the results from sampling coherent spin states to converge at least as quickly.

This method dramatically improves upon a deterministic evaluation of the ensemble average of an observable by reducing the number of wavepackets which must be propagated. For the model radical pair discussed here, it is over 5000 times faster, corresponding to one day of CPU time rather than 13 years. However, the computational time required for each wavepacket propagation still scales as $\mathcal{O}(Z \log Z)$, meaning calculations of radical pairs will still become intractable as N increases. In order to overcome this, we have developed two semiclassical models which describe radical pair reactions, which will be introduced in Chap. 3.

2.6 Spin Correlation Tensors

For the special case of radical pairs with no electron spin coupling ($\mathbf{D}' = 0$) and whose singlet and triplet recombination rates are equal ($k_\mathrm{S} = k_\mathrm{T} = k$), there is another way to reduce the computational time required to calculate the singlet yield. In this case, the ensemble average of the singlet probability of can be expressed as a function of the spin correlation tensors of each radical [30].

Beginning with the definition of the ensemble average from Eq. (2.38),

$$\mathrm{P_S}(t) = \mathrm{tr}\left[\hat{\rho}(t)\hat{P}_\mathrm{S}\right], \tag{2.63}$$

inserting Eq. (2.41) gives

$$\mathrm{P_S}(t) = \mathrm{tr}\left[e^{-i\hat{H}t-\hat{K}t}\hat{\rho}(0)e^{+i\hat{H}t-\hat{K}t}\hat{P}_\mathrm{S}\right]. \tag{2.64}$$

Since in this example $\hat{K} = {}^{k}/_{2}\hat{1}$, we can immediately factor out the recombination operator

$$\mathrm{P_S}(t) = e^{-kt}\mathrm{tr}\left[e^{-i\hat{H}t}\hat{\rho}(0)e^{+i\hat{H}t}\hat{P}_\mathrm{S}\right], \tag{2.65}$$

and we see that the spin dynamics of the radical pair are decoupled from its recombination. Considering the case of a radical pair formed by photoexcitation in the singlet state, such that $\rho(0) = \hat{P}_\mathrm{S}/Z$, and using the invariance of a trace to cyclic permutation

$$\mathrm{P_S}(t) = \frac{e^{-kt}}{Z}\mathrm{tr}\left[\hat{P}_\mathrm{S}e^{+i\hat{H}t}\hat{P}_\mathrm{S}e^{-i\hat{H}t}\right]. \tag{2.66}$$

Inserting Eq. (2.32) yields

$$\begin{aligned}
\mathrm{P_S}(t) &= \frac{e^{-kt}}{Z}\mathrm{tr}\left[\left(\frac{1}{4}\hat{1} - \hat{\mathbf{S}}_1\cdot\hat{\mathbf{S}}_2\right)e^{+i\hat{H}t}\left(\frac{1}{4}\hat{1} - \hat{\mathbf{S}}_1\cdot\hat{\mathbf{S}}_2\right)e^{-i\hat{H}t}\right] \\
&= \frac{e^{-kt}}{Z}\mathrm{tr}\left[\frac{1}{16}\hat{1} - \frac{1}{2}\hat{\mathbf{S}}_1\cdot\hat{\mathbf{S}}_2 + \hat{\mathbf{S}}_1\cdot\hat{\mathbf{S}}_2\, e^{+i\hat{H}t}\hat{\mathbf{S}}_1\cdot\hat{\mathbf{S}}_2\, e^{-i\hat{H}t}\right] \\
&= \frac{e^{-kt}}{Z}\left(\mathrm{tr}\left[\frac{1}{2}\hat{P}_\mathrm{S} - \frac{1}{16}\hat{1}\right] + \mathrm{tr}\left[\hat{\mathbf{S}}_1\cdot\hat{\mathbf{S}}_2\, e^{+i\hat{H}t}\hat{\mathbf{S}}_1\cdot\hat{\mathbf{S}}_2\, e^{-i\hat{H}t}\right]\right).
\end{aligned} \tag{2.67}$$

The first term of the final line of Eq. (2.67) is simply a constant. The trace of the unit operator returns the number of states in the total Hilbert space, and the trace of the singlet projection operator the number of states in the singlet subspace, $4Z$ and Z respectively. The second term in Eq. (2.67) may be expanded by noting that in this case $\hat{H} = \hat{H}_1 + \hat{H}_2$, and that $\hat{H}_1$ and $\hat{H}_2$ commute,

$$\begin{aligned}
\hat{\mathbf{S}}_1\cdot\hat{\mathbf{S}}_2\, e^{+i\hat{H}t}\hat{\mathbf{S}}_1\cdot\hat{\mathbf{S}}_2\, e^{-i\hat{H}t} &= \sum_{\alpha,\beta} \hat{S}_{1\alpha}\hat{S}_{2\alpha}\, e^{+i(\hat{H}_1+\hat{H}_2)t}\,\hat{S}_{1\beta}\hat{S}_{2\beta}\, e^{-i(\hat{H}_1+\hat{H}_2)t} \\
&= \sum_{\alpha,\beta} \hat{S}_{1\alpha}\, e^{+i\hat{H}_1 t}\,\hat{S}_{1\beta}\, e^{-i\hat{H}_1 t}\,\hat{S}_{2\alpha}\, e^{+i\hat{H}_2 t}\,\hat{S}_{2\beta}\, e^{-i\hat{H}_2 t}.
\end{aligned} \tag{2.68}$$

Finally, defining the spin correlation tensor

$$R^{(i)}_{\alpha\beta}(t) = \frac{1}{Z_i}\mathrm{tr}\left[\hat{S}_{i\alpha}\, e^{+i\hat{H}_i t}\,\hat{S}_{i\beta}\, e^{-i\hat{H}_i t}\right], \tag{2.69}$$

where Z_i is the number of nuclear spin states of radical i, and combining Eqs. (2.67), (2.68) and (2.69), we obtain

$$\mathrm{P_S}(t) = e^{-kt}\left(\frac{1}{4} + \sum_{\alpha,\beta} R^{(1)}_{\alpha\beta}(t) R^{(2)}_{\alpha\beta}(t)\right). \tag{2.70}$$

While at first sight this may seem to have complicated the problem, in fact this formulation of $\mathrm{P_S}(t)$ greatly reduces the time required for its calculation. The correlation tensors in Eq. (2.70) are independent; they may be calculated separately in the reduced Hilbert space of each individual radical, which are of size $2Z_i$. This will be much faster than a calculation in the total Hilbert space of size $4Z_1Z_2 = 4Z$. However, Eq. (2.70) only applies to the specific case of a radical pair with uncoupled electrons and symmetric recombination rates. These circumstances rarely arise in real systems; instead, this formulation will be useful when performing calculations of model systems for comparison with the semiclassical approximations of Chap. 3.

2.7 Relaxation

Radical pairs are formed with electron spins in a correlated, non-equilibrium state. As they interact with their environment, they relax towards equilibrium, in which all correlations and coherences are lost and the populations of the singlet and triplet states are given by the Boltzmann distribution [10, 12, 31]. Spin relaxation is driven in general by the time-dependent modulation of magnetic interactions. However, since these modulations are typically induced by molecular motions, the exact mechanisms of relaxation will vary from system to system, as different motions modulate different interactions in each species. These motions make the explicit quantum mechanical treatment of relaxation extremely complex.

In this section, we will discuss the two relaxation mechanisms which are of particular relevance to radical pair reactions. However, even simplified models of these relaxation mechanisms present a computational challenge. The quantum mechanics of these processes necessitate a description of the radical pairs in Liouville space [32, 33], whose size is $(4Z)^2$. As a result, the time required for calculations that include relaxation becomes prohibitive even for radical pairs with around ten nuclear spins, further limiting the applicability of fully quantum mechanical simulations. This provides a further motivation for developing semiclassical theories of radical pair reactions, which can account for these phenomena in an approximate but very efficient way, as discussed in Sect. 3.4.

2.7.1 *Modulation of Hyperfine Tensors*

For radical pairs in solution, the most important relaxation mechanism is usually the modulation of anisotropic hyperfine tensors by molecular tumbling. The simplest way to describe this relaxation is to use the phenomenological Bloch equations [34]:

$$\begin{aligned}\frac{\mathrm{d}\langle\hat{S}_z\rangle}{\mathrm{d}t} &= -\frac{\langle\hat{S}_z\rangle - \langle\hat{S}_{eq}\rangle}{T_1},\\ \frac{\mathrm{d}\langle\hat{S}_x\rangle}{\mathrm{d}t} &= -\frac{\langle\hat{S}_x\rangle}{T_2},\\ \frac{\mathrm{d}\langle\hat{S}_y\rangle}{\mathrm{d}t} &= -\frac{\langle\hat{S}_y\rangle}{T_2},\end{aligned} \tag{2.71}$$

where $\langle\hat{S}_q\rangle$ is the expectation value of the q component of the electron spin operator, and $\langle\hat{S}_{eq}\rangle$ is the equilibrium value of $\langle\hat{S}_z\rangle$ when an external field is applied in the z direction. Since changes in $\langle\hat{S}_z\rangle$ alter the total Zeeman interaction energy, only components of the modulating hyperfine interactions which involve energy transfer with the surroundings may cause longitudinal relaxation. However, changes in $\langle\hat{S}_x\rangle$ and $\langle\hat{S}_y\rangle$ do not change the energy of the system, and so may be caused by different components of the hyperfine tensor modulation. As a result, the two relaxation times T_1 and T_2 are usually different [12]. While the simple exponential model predicted by Eqs. (2.71) is rarely justified [10], it is a very straightforward way to model spin relaxation which provides qualitatively useful and intuitive results, and so remains quite widely used.

A more complex perturbative approach to describing relaxation was developed by Redfield [32, 35]. In this theory, the system is perturbed by a time dependent Hamiltonian, $\hat{H}_1(t)$, which arises from some random motion and whose time average is zero. The resulting relaxation of the density operator can be shown to be described by a linear set of coupled differential equations [32]:

$$\frac{\mathrm{d}\hat{\rho}_{\alpha\alpha'}}{\mathrm{d}t} = +\sum_{\beta\beta'} R_{\alpha\alpha'\beta\beta'}\hat{\rho}_{\beta\beta'} \tag{2.72}$$

where $\hat{\rho}_{\alpha\alpha'}$ is an element of the density operator. $R_{\alpha\alpha'\beta\beta'}$ is an element of the relaxation superoperator, which describes the evolution of each element of the density matrix under the perturbing Hamiltonian. This result only holds if the motion accounted for by $\hat{H}_1(t)$ is Marvokian – that is, the correlation time of the motion is sufficiently short that the evolution of $\hat{\rho}(t)$ depends only on its current value, and not its previous values [10]. Redfield theory itself is completely general, but in the context of radical pair reactions it is typically applied by defining $\hat{H}_1(t)$ to describe how a particular molecular motion of interest causes fluctuations of the anisotropic hyperfine interactions [36, 37].

For certain choices of $\hat{H}_1(t)$, Redfield theory reduces to the Bloch equations (2.71) and provides expressions for the T_1 and T_2 relaxation times in terms of the spectral density of the perturbation [12]. In general, however, it leads to an extremely complex expression for relaxation, in which every element of the density matrix is coupled to every other element, such that the relaxation cannot be characterised by a single relaxation time. As a result, in practice it is applicable only to very small systems.

2.7.2 *Singlet-Triplet Dephasing*

Singlet-triplet dephasing is a relaxation mechanism which results solely in the decay of coherences between the singlet and triplet states, and does not change the populations of those states. It is produced by modulation of the exchange interaction between the spins in the radical pair. This may be caused by one of the electrons in the pair hopping to a third site to temporarily form a different radical pair with a much larger exchange coupling, before hopping back to reform the original radical pair. Dephasing by this mechanism was originally proposed to explain the observation that the line splittings in the electron spin resonance (ESR) [38] and stimulated nuclear polarization (SNP) [39] spectra of some radical pairs were half the value of the splitting measured in the corresponding free radical [33]. More recently, it has been invoked to explain the fact the values of $B_{1/2}$ observed in the cryptochrome family are much larger than those predicted by the Weller equation [40].

In Ref. [33], Shushin shows that provided the lifetime of the original radical pair is much longer than that of the intermediate radical pair, there is no need to explicitly include the third site in the description of the system. Instead, the singlet-triplet dephasing may be treated phenomenologically using a Lindblad equation of the form:

$$\frac{d\hat{\rho}}{dt} = \cdots + k_{\mathrm{R}} \sum_{j=\mathrm{S,T}} \left(\hat{P}_j \hat{\rho} \hat{P}_j - \frac{1}{2} \hat{P}_j^{\dagger} \hat{P}_j \hat{\rho} - \frac{1}{2} \hat{\rho} \hat{P}_j^{\dagger} \hat{P}_j \right). \tag{2.73}$$

Using the property of projection operators that $\hat{P}_j^{\dagger} \hat{P}_j = \hat{P}_j$, and the fact that $\hat{P}_{\mathrm{S}} + \hat{P}_{\mathrm{T}} = \hat{1}$, this expression can be simplified:

$$\begin{aligned}
\frac{d\hat{\rho}}{dt} &= \cdots + k_{\mathrm{R}} \left(\hat{P}_{\mathrm{S}} \hat{\rho} \hat{P}_{\mathrm{S}} - \frac{1}{2} \hat{P}_{\mathrm{S}} \hat{\rho} - \frac{1}{2} \hat{\rho} \hat{P}_{\mathrm{S}} + \hat{P}_{\mathrm{T}} \hat{\rho} \hat{P}_{\mathrm{T}} - \frac{1}{2} \hat{P}_{\mathrm{T}} \hat{\rho} - \frac{1}{2} \hat{\rho} \hat{P}_{\mathrm{T}} \right) \\
&= \cdots + k_{\mathrm{R}} \left(\hat{P}_{\mathrm{S}} \hat{\rho} \hat{P}_{\mathrm{S}} + \hat{P}_{\mathrm{T}} \hat{\rho} \hat{P}_{\mathrm{T}} - \frac{1}{2} \left(\hat{P}_{\mathrm{S}} + \hat{P}_{\mathrm{T}} \right) \hat{\rho} - \frac{1}{2} \hat{\rho} \left(\hat{P}_{\mathrm{S}} + \hat{P}_{\mathrm{T}} \right) \right) \\
&= \cdots + k_{\mathrm{R}} \left(\hat{P}_{\mathrm{S}} \hat{\rho} \hat{P}_{\mathrm{S}} + \hat{P}_{\mathrm{T}} \hat{\rho} \hat{P}_{\mathrm{T}} - \left(\hat{P}_{\mathrm{S}} + \hat{P}_{\mathrm{T}} \right) \hat{\rho} \left(\hat{P}_{\mathrm{S}} + \hat{P}_{\mathrm{T}} \right) \right) \\
&= \cdots - k_{\mathrm{R}} \left(\hat{P}_{\mathrm{S}} \hat{\rho} \hat{P}_{\mathrm{T}} + \hat{P}_{\mathrm{T}} \hat{\rho} \hat{P}_{\mathrm{S}} \right).
\end{aligned} \tag{2.74}$$

It is clear from the final line of Eq. (2.74) that this approach will result in the decay of coherences between the singlet and triplet states at a rate k_{R}, while leaving their populations unchanged, as required.

2.8 Conclusion

In this chapter, we have introduced the quantum mechanical operators required to describe radical pair reactions. We have outlined the origin of each term of the spin Hamiltonian in Eq. (2.1), and considered two possible forms of the recombination operator which accounts for the reaction of radical pairs. We have then demonstrated how the ensemble averages of observables may be efficiently calculated in order to make comparisons between theory and experiment. The method outlined in Sect. 2.5 will be used in Chap. 4 to calculate the triplet yields and survival probabilities of a series of molecular wires. Using this information, we shall investigate the mechanisms of recombination which operate along the wires. Finally, we discussed how the computational demands of a fully quantum mechanical simulation motivate the development of approximate semiclassical theories. In Chap. 3 these semiclassical models will be presented and the results they produce compared to exact results obtained using the methods outlined here.

References

1. McLauchlan, K. A., & Steiner, U. E. (1991). The spin-correlated radical pair as a reaction intermediate. *Molecular Physics*, *73*, 241–263.
2. Steiner, U. E., & Ulrich, T. (1989). Magnetic field effects in chemical kinetics and related phenomena. *Chemical Reviews*, *89*, 51–147.
3. Pauli, W. (1940). The connection between spin and statistics. *Physical Review*, *58*, 716–722.
4. Van Vleck, J. H. (1932). *The Theory of Electric and Magnetic Susceptibilities*. Oxford: Clarendon Press.
5. Dirac, P. A. M. (1930). *The Principles of Quantum Mechanics*. Oxford: Clarendon Press.
6. Atkins, P. W., & Friedman, R. S. (2011). *Molecular Quantum Mechanics*. Oxford: Oxford University Press.
7. Anderson, P. W. (1950). Antiferromagnetism. *Theory of Superexchange Interaction. Physical Review*, *79*, 350–356.
8. Anderson, P. W. (1959). New approach to the theory of superexchange interactions. *Physical Review*, *115*, 2–13.
9. McConnell, H. M. (1961). Intramolecular charge transfer in aromatic free radicals. *Journal of Chemical Physics*, *35*, 508–515.
10. Abragam, A. (1961). *The Principles of Nuclear Magnetism*. Oxford: Clarendon Press.
11. Jackson, J. D. (1975). *Classical Electrodynamics*. New York: Wiley.
12. Slichter, C. P. (1963). *Principles of Magnetic Resonance*. Berlin: Springer.
13. Carrington, A., & McLachlan, A. D. (1967). *Introduction to Magnetic Resonance with Applications to Chemistry and Chemical Physics*. New York: Harper & Row.
14. Rieger, P. (2007). *Electron Spin Resonance*. Cambridge: The Royal Society of Chemistry.
15. McConnell, H. M., & Chesnut, D. B. (1958). Theory of isotropic hyperfine interactions in ϕ-electron radicals. *Journal of Chemical Physics*, *28*, 107–117.
16. Haberkorn, R. (1976). Density matrix description of spin-selective radical pair reactions. *Molecular Physics*, *32*, 1491–1493.
17. Kominis, I. K. (2009). Quantum Zeno effect explains magnetic-sensitive radical-ion-pair reactions. *Physical Review E*, *80*, 056115.
18. Maeda, K., et al. (2008). Chemical compass model of avian magnetoreception. *Nature*, *453*, 387–390.
19. Jones, J. A., & Hore, P. J. (2010). Spin-selective reactions of radical pairs act as quantum measurements. *Chemical Physics Letters*, *488*, 90–93.
20. Jones, J. A., Maeda, K., & Hore, P. J. (2011). Reaction operators for spin-selective chemical reactions of radical pairs. *Chemical Physics Letters*, *507*, 269–273.
21. Maeda, K., Liddell, P. A., Gust, D., & Hore, P. J. (2013). Spin-selective recombination reactions of radical pairs: Experimental test of validity of reaction operators. *Journal of Chemical Physics*, *139*, 234309.
22. Ivanov, K. L., Petrova, M. V., Lukzen, N. N., & Maeda, K. (2010). Consistent treatment of spin-selective recombination of a radical pair confirms the Haberkorn approach. *Journal of Physical Chemistry A*, *114*, 9447–9455.
23. Sakurai, J. J. (1986). *Modern Quantum Mechanics*. Pearson.
24. Von Neumann, J. (1955). *Mathematical Foundations of Quantum Mechanics*. Princeton: Princeton University Press.
25. Lewis, A. M., Manolopoulos, D. E., & Hore, P. J. (2014). Asymmetric recombination and electron spin relaxation in the semiclassical theory of radical pair reactions. *Journal of Chemical Physics*, *141*, 044111.
26. Radclffie, J. M. (1971). Some properties of coherent spin states. *Journal of Physics A: General Physics*, *4*, 313–323.
27. Lieb, E. H. (1973). The classical limit of quantum spin systems. *Communications in Mathematical Physics*, *31*, 327–340.
28. Zhang, W.-M., Feng, D. H., & Gilmore, R. (1990). Coherent states: theory and some applications. *Reviews of Modern Physics*, *62*, 867–927.

29. Hammersley, J. M., & Handscomb, D. C. (1964). *Monte Carlo Methods 50–75*. Netherlands: Springer.
30. Schulten, K., & Wolynes, P. G. (1978). Semi-classical description of electron-spin motion in radicals including effect of electron hopping. *Journal of Chemical Physics*, *68*, 3292–3297.
31. Atherton, N. M. (1973). *Electron Spin Resonance: Theory and Applications*. S.l.: Ellis Horwood Limited.
32. Redfield, A. G. (1957). On the theory of relaxation processes. *IBM Journal of Research Development*, 19–31 .
33. Shushin, A. I. (1991). The effect of the spin exchange interaction on SNP and RYDMR spectra of geminate radical pairs. *Chemical Physics Letters*, *181*, 274–278.
34. Bloch, F. (1946). Nuclear induction. *Physical Review*, *70*, 460–474.
35. Redfield, A. G. (1965). The theory of relaxation processes. *Advances in Magnetic and Optical Resonance*, *1*, 1–32.
36. Shushin, A. I. (2011). Magnetic field effects on electron-hole recombination in disordered organic semiconductors. *Physical Review B*, *84*, 115212.
37. Worster, S., Kattnig, D. R., & Hore, P. J. (2016). Spin relaxation of radicals in cryptochrome and its role in avian magnetoreception. *Journal of Chemical Physics*, *145*, 035104.
38. Closs, G. L., & Forbes, M. D. E. (1987). Observation of medium chain length Polymethylene Biradicals in liquid solutions by time resolved EPR spectroscopy. *Journal of the American Chemical Society*, *109*, 6185–6187.
39. Koptyug, I. V., Bagryanskaya, E. G., & Sagdeev, R. Z. (1989). Observation of 13C SNP during the photolysis of cyclododecanone in solution. *Chemical Physics Letters*, *163*, 503–508.
40. Maeda, K., et al. (2012). Magnetically sensitive light-induced reactions in cryptochrome are consistent with its proposed role as a magnetoreceptor. *Proceedings of the National Academy of Sciences of the United States of America*, *109*, 4774–4779.

Chapter 3
Semiclassical Approximations

In Chap. 2, we outlined the quantum mechanics which describes radical pair reactions, and demonstrated an efficient way to perform fully quantum mechanical simulations. However, the computational time required for these calculations scales exponentially with the number of nuclear spins in the radical pair, meaning there is a limit to the size of radical pair which may practically be treated in this way. Furthermore, including spin relaxation in quantum mechanical simulations is an extremely complex task. In order to avoid the scaling problem, and to allow a straightforward phenomenological account of relaxation, we have considered two semiclassical models of radical pair reactions, which will be presented in this chapter.[1]

In 1978, Schulten and Wolynes presented their semiclassical theory of electron spin dynamics in radical pairs [1]. It was based on the approximation of quantum mechanical vector operators as classical vectors, and then further assumed that the nuclear spins were static on the time-scale of the electron spin motion. In certain limiting cases this theory gives analytical results; even in the most general case it produces a set of 16 coupled differential equations which may be easily solved numerically. It is computationally a huge improvement over the exponential scaling of quantum mechanics, and for certain radical pairs like pyrene-N,N,-dimethylalanine produces very accurate results [1]. However, as we will demonstrate in Sect. 3.3, it does not always agree well with exact quantum mechanical results, particularly at low applied magnetic field strengths.

Inspired by this approach, we have recently developed an improved semiclassical theory which produces accurate results for a wider range of radical pairs [2, 3]. In this theory, the electron and nuclear spins are still treated as classical vectors, but the assumption that the motion of the nuclear spins can be neglected is removed – the nuclear spins are allowed to precess around the electron spin. While this does make

[1] Parts of this chapter have been reproduced with permission from Lewis, A. M., Manolopoulos, D. E. and Hore, P. J. *Journal of Chemical Physics* **141**, 044111 (2014), https://aip.scitation.org/doi/abs/10.1063/1.4890659.

A. Lewis, *Spin Dynamics in Radical Pairs*, Springer Theses,
https://doi.org/10.1007/978-3-030-00686-0_3

the improved semiclassical method computationally more expensive than Schulten and Wolynes' theory, the CPU time required for these simulations scales only linearly with the number of nuclear spins in the system, avoiding the exponential scaling of quantum mechanical calculations. We will demonstrate in Sect. 3.3 that this approach produces more accurate results than Schulten and Wolynes' method.

In this chapter we will outline the assumptions which lead to each semiclassical theory. While Schulten–Wolynes (SW) theory was developed first, for reasons of clarity we will present the improved semiclassical (SC) theory first, and then describe the additional approximations which lead to SW theory. We will then compare the results produced by the two semiclassical methods to exact quantum mechanical results for a number of different model problems. Finally, we will briefly discuss how relaxation of the electron spins can be accounted for the in the SC theory.

3.1 The Semiclassical Theory

The semiclassical theory is based on the approximation that the electron and nuclear spins may be treated as classical vectors. These classical vectors evolve under a set of coupled equations of motion which may be solved numerically; the number of coupled equations will increase linearly with the number of nuclear spins. The equations of motion may be obtained by deriving the Heisenberg equations of motion for the quantum mechanical electron and nuclear spin operators, and then simply replacing all of the vector operators with classical vectors. For simplicity, throughout this section we will neglect both exchange and dipolar electron spin coupling, setting $\mathbf{D}' = 0$, and assume that all hyperfine interactions are isotropic, $\mathbf{A}' = a\mathbf{E}$, working with a Hamiltonian of the form

$$\hat{H} = \boldsymbol{\omega}_1 \cdot \hat{\mathbf{S}}_1 + \boldsymbol{\omega}_2 \cdot \hat{\mathbf{S}}_2 + \sum_{k=1}^{N_1} a_{1k}\hat{\mathbf{I}}_{1k} \cdot \hat{\mathbf{S}}_1 + \sum_{k=1}^{N_2} a_{2k}\hat{\mathbf{I}}_{2k} \cdot \hat{\mathbf{S}}_2. \tag{3.1}$$

The additional terms in the equations of motion which arise from using the full Hamiltonian in Eq. (2.1) are given in Appendix C; their inclusion does not change any of the following discussion.

The Heisenberg equation of motion for an operator $\hat{A}(t)$ defined by Eq. (2.42) is

$$\begin{aligned} \frac{\mathrm{d}}{\mathrm{d}t}\hat{A}(t) &= \frac{\mathrm{d}}{\mathrm{d}t}\left(e^{+i\hat{H}t-\hat{K}t}\,\hat{A}\,e^{-i\hat{H}t-\hat{K}t}\right) \\ &= \left[\left(+i\hat{H}-\hat{K}\right)\hat{A}(t) + \hat{A}(t)\left(-i\hat{H}-\hat{K}\right)\right] \\ &= +i\left[\hat{H},\hat{A}(t)\right] - \left\{\hat{K},\hat{A}(t)\right\}. \end{aligned} \tag{3.2}$$

There are a total of 16 operators in the complete set of variables which describe the two electron spins of the radical pair: $\hat{\mathbf{S}}_1$, $\hat{\mathbf{S}}_2$, $\hat{\mathbf{T}}_{12} = \hat{\mathbf{S}}_1 \otimes \hat{\mathbf{S}}_2$ and $\hat{1}$. Firstly, we must derive the Heisenberg equation of motion for each of these operators. Then, we replace the vector operators with classical vectors, and close the equations of motion by replacing the operator $\hat{\mathbf{S}}_i \times \hat{\mathbf{I}}_{ik}$ wherever it appears with the cross product of the classical vectors $\mathbf{S}_i$ and $\mathbf{I}_{ik}$. Doing this, we find a set of $3N + 16$ coupled equations [3]:

$$\tfrac{d}{dt}\mathbf{I}_{ik} = a_{ik}\mathbf{S}_i \times \mathbf{I}_{ik} - \bar{k}\,\mathbf{I}_{ik} + 4\Delta k\,\mathrm{tr}[\mathbf{T}_{12}]\mathbf{I}_{ik}, \tag{3.3}$$

$$\tfrac{d}{dt}\mathbf{S}_1 = \bar{\boldsymbol{\omega}}_1 \times \mathbf{S}_1 - \bar{k}\,\mathbf{S}_1 + \Delta k\,\mathbf{S}_2, \tag{3.4}$$

$$\tfrac{d}{dt}\mathbf{S}_2 = \bar{\boldsymbol{\omega}}_2 \times \mathbf{S}_2 - \bar{k}\,\mathbf{S}_2 + \Delta k\,\mathbf{S}_1, \tag{3.5}$$

$$\tfrac{d}{dt}\mathbf{T}_{12} = \bar{\boldsymbol{\omega}}_1 \times \mathbf{T}_{12} - \mathbf{T}_{12} \times \bar{\boldsymbol{\omega}}_2 - \bar{k}\mathbf{T}_{12} + \Delta k\,\mathbf{T}_{12}^{\top} + \Delta k\,\bar{P}_{\mathrm{S}}\,\mathbf{E}, \tag{3.6}$$

and

$$\frac{d}{dt}\bar{\mathbf{1}} = -\bar{k}\,\bar{\mathbf{1}} + 4\Delta k\,\mathrm{tr}[\mathbf{T}_{12}], \tag{3.7}$$

where $\bar{\mathbf{1}}$ is the classical variable corresponding to the unit operator, and

$$\bar{k} = \frac{1}{4}(k_{\mathrm{S}} + 3k_{\mathrm{T}}), \tag{3.8}$$

$$\Delta k = \frac{1}{4}(k_{\mathrm{S}} - k_{\mathrm{T}}), \tag{3.9}$$

$$\bar{\boldsymbol{\omega}}_i = \boldsymbol{\omega}_i + \sum_{k=1}^{N_i} a_{ik}\,\mathbf{I}_{ik}. \tag{3.10}$$

In Eqs. (3.3) and (3.7), $\mathrm{tr}[\mathbf{T}_{12}]$ denotes the trace of the tensor $\mathbf{T}_{12}$, and is the classical variable corresponding to the operator $\hat{\mathbf{S}}_1 \cdot \hat{\mathbf{S}}_2$. In Eq. (3.6), $\bar{\boldsymbol{\omega}}_1 \times \mathbf{T}_{12}$ denotes the vector product of $\bar{\boldsymbol{\omega}}_1$ with each column of $\mathbf{T}_{12}$, $\mathbf{T}_{12} \times \bar{\boldsymbol{\omega}}_2$ denotes the vector product of each row of $\mathbf{T}_{12}$ with $\bar{\boldsymbol{\omega}}_2$, $\mathbf{T}_{12}^{\top}$ is the transpose of $\mathbf{T}_{12}$, $\mathbf{E}$ is a (3×3) unit matrix, and

$$\bar{P}_{\mathrm{S}} = \frac{1}{4}\bar{\mathbf{1}} - \mathrm{tr}[\mathbf{T}_{12}] \tag{3.11}$$

is the classical variable corresponding to the singlet projection operator $\hat{P}_{\mathrm{S}}$ in Eq. (2.32). Similarly, the classical variable corresponding to the triplet projection operator is

$$\bar{P}_{\mathrm{T}} = \frac{3}{4}\bar{\mathbf{1}} + \mathrm{tr}[\mathbf{T}_{12}]. \tag{3.12}$$

An alternative way of expressing the equations of motion in Eqs. (3.3)–(3.7) using tensor notation and the Einstein summation convention is given in Appendix C.

The initial conditions of these classical variables are specified as follows. We replace the trace over all of the spin states of each electron and nuclear spin with the integral over all possible orientations of the corresponding classical vector,

$$\begin{aligned}\mathrm{tr}_{S_i} &\rightarrow \frac{2S_i+1}{4\pi}\int d\Omega_{S_i},\\ \mathrm{tr}_{I_{ik}} &\rightarrow \frac{2I_{ik}+1}{4\pi}\int d\Omega_{I_{ik}}.\end{aligned} \tag{3.13}$$

The two-electron classical variables $\mathbf{T}_{12}$ are simply the outer product of the classical electron spin vectors $\mathbf{S}_1 \otimes \mathbf{S}_2$ at $t = 0$, although this will not in general be true at later times. $\bar{\mathbf{1}}$ is equal to unity at t = 0, but in general will change with time to account for the recombination of the radical pair. Therefore, the semiclassical expression analogous to the quantum mechanical Eq. (2.43) for the ensemble average of an observable $\mathrm{A}(t)$ of a radical pair is

$$\mathrm{A}(t) \simeq \prod_{i=1}^{2} \frac{2}{4\pi}\int d\Omega_{S_i} \prod_{k=1}^{N_i} \frac{2I_{ik}+1}{4\pi}\int d\Omega_{I_{ik}}\, \bar{\rho}(0)\bar{A}(t), \tag{3.14}$$

where $\bar{\rho}(0)$ and $\bar{A}(t)$ are the classical variables corresponding to the operators $\hat{\rho}(0)$ and $\hat{A}(t)$ respectively. Similarly, the semiclassical expression analogous to Eq. (2.69) for the spin correlation tensor of a radical in a singlet-born pair is

$$R^{(i)}_{\alpha\beta}(t) \simeq \frac{2}{4\pi}\int d\Omega_{S_i} \prod_{k=1}^{N_i} \frac{1}{4\pi}\int d\Omega_{I_{ik}}\, S_{i\alpha}(0)S_{i\beta}(t). \tag{3.15}$$

Note that there are no factors of $(2I_{ik}+1)$ in Eq. (3.15) because they cancel with the prefactor $1/Z_i$ in Eq. (2.69).

However, an examination of Eqs. (3.3)–(3.7) reveals a problem with the equations of motion for the classical variables, which can be clearly seen if we consider the scenario where $\Delta k = 0$ and $\bar{k} \neq 0$. As discussed in Sect. 2.6, under these conditions the recombination is separable from the spin dynamics of the radical pair. That is, in the exact QM formulation we can calculate the spin dynamics of the radical pair in the absence of any recombination and subsequently multiply the resulting curves by an exponential decay to account for the recombination of the radical pair to obtain results which are identical to those calculated without factoring out recombination. The equations of motion in Eqs. (3.3)–(3.7) are not consistent with this. Furthermore, in Eq. (3.3), each nuclear spin vector $\mathbf{I}_{ik}$ precesses around the electron spin vector $\mathbf{S}_i$, which has length $|\mathbf{S}_i|$. However, Eqs. (3.4) and (3.5) show that the magnitudes of the electron spin vectors will decrease with time, slowing the precession of the nuclear spin vectors. This is unphysical – there is no reason why the nuclear spin precession should slow down before the radical pair recombines, at which point any precession would cease entirely.

Clearly some correction to the equations of motion is required. Before we consider what that correction might be, though, it is helpful to consider another special case. We can show numerically that our semiclassical theory becomes exact in the absence of any nuclear spins, such that $\bar{\omega}_i = \omega_i$ is constant. Figures 3.1 and 3.2 show the singlet and triplet probability and the xy component of the spin correlation tensors

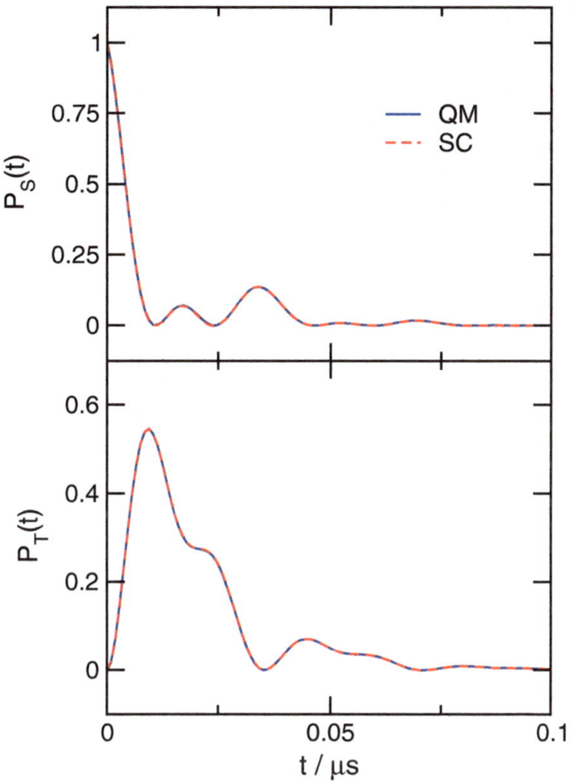

Fig. 3.1 Comparison of the quantum mechanical (QM) and semiclassical (SC) singlet and triplet probabilities of a radical pair with $\bar{\omega}_1 = \omega_1 = (0, 0, 1)$, $\bar{\omega}_2 = \omega_2 = -(\sqrt{1/2}, \sqrt{1/3}, \sqrt{1/5})$, $k_S = \sqrt{1/7}$, and $k_T = \sqrt{1/11}$. Adapted from Ref. [3]

of each radical respectively in a radical pair with no nuclear spins, asymmetric recombination rates and $\omega_1 \neq \omega_2$. The spin correlation tensors are functions of the single electron spin variables, while the singlet and triplet probabilities depend on the two-electron variables and $\bar{\mathbf{1}}$; the two sets of variables are independent of one another when $\bar{\omega}_1$ and $\bar{\omega}_2$ are constant and there is no electron spin coupling. Therefore, Fig. 3.1 demonstrates that Eqs. (3.6) and (3.7) are exact and Fig. 3.2 shows that Eqs. (3.4) and (3.5) are exact in this case, so none of these equations should be modified when we come to correct our semiclassical theory.[2]

In addition, Eq. (3.7) satisfies the requirement that the sum of the singlet and triplet yield equals one precisely. This is straightforward to prove quantum mechanically [3]. The singlet and triplet yields are defined as

$$\Phi_S = k_S \int_0^\infty P_S(t)\, dt \tag{3.16}$$

[2]These equations remain exact when electron spin coupling is included, as shown numerically in Appendix C.

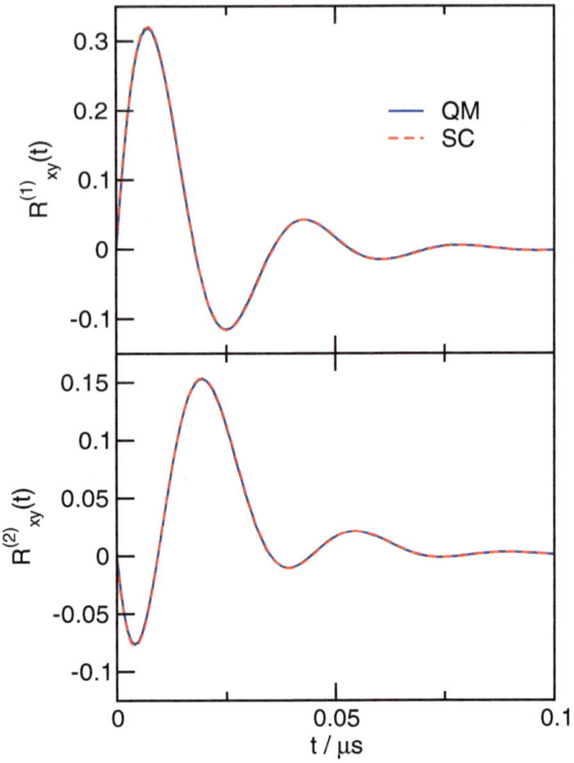

Fig. 3.2 Comparison of the quantum mechanical (QM) and semiclassical (SC) xy components of the spin correlation tensors of the electrons in radicals 1 and 2 of a radical pair with $\bar{\omega}_1 = \omega_1 = (0, 0, 1)$, $\bar{\omega}_2 = \omega_2 = -(\sqrt{1/2}, \sqrt{1/3}, \sqrt{1/5})$, $k_S = \sqrt{1/7}$, and $k_T = \sqrt{1/11}$. Adapted from Ref. [3]

and

$$\Phi_T = k_T \int_0^\infty P_T(t)\, dt \tag{3.17}$$

respectively. Inserting the definition of the ensemble average of an observable from Eq. (2.43), we can show that [3]

$$\begin{aligned}
\Phi_S + \Phi_T &= \mathrm{tr}\left\{\hat{\rho}(0) \int_0^\infty \left(k_S \hat{P}_S(t) + k_T \hat{P}_T(t)\right) dt\right\} \\
&= \mathrm{tr}\left\{\hat{\rho}(0) \int_0^\infty 2\hat{K}(t)\, dt\right\} \\
&= \mathrm{tr}\left\{\hat{\rho}(0) \int_0^\infty e^{+i\hat{H}t-\hat{K}t}\, 2\hat{K}\, e^{-i\hat{H}t-\hat{K}t}\, dt\right\} \\
&= \mathrm{tr}\left\{\hat{\rho}(0) \int_0^\infty \frac{d}{dt}\left(-e^{+i\hat{H}t-\hat{K}t}\, \hat{1}\, e^{-i\hat{H}t-\hat{K}t}\right) dt\right\} \\
&= \mathrm{tr}\left\{\hat{\rho}(0) \int_0^\infty \frac{d}{dt}\left(-\hat{1}(t)\right) dt\right\} \\
&= \mathrm{tr}\left\{\hat{\rho}(0) \left[-e^{+i\hat{H}t-\hat{K}t} e^{-i\hat{H}t-\hat{K}t}\right]_0^\infty\right\} \\
&= \mathrm{tr}\left\{\hat{\rho}(0)\right\} = 1.
\end{aligned} \tag{3.18}$$

We can use a similar argument to show that our semiclassical theory also satisfies this condition. Using the semiclassical definitions of the singlet and triplet probabilities in Eqs. (3.11) and (3.12), we can show that by analogy to the quantum mechanics above

$$\begin{aligned} k_S \bar{P}_S(t) + k_T \bar{P}_T(t) &= \frac{1}{4}(k_S + 3k_T)\,\bar{\mathbf{1}}(t) + (k_T - k_S)\,\mathrm{tr}[\mathbf{T}_{12}(t)] \\ &= \bar{k}\bar{\mathbf{1}}(t) - 4\Delta k\,\mathrm{tr}[\mathbf{T}_{12}(t)] \\ &= -\frac{\mathrm{d}}{\mathrm{d}t}\bar{\mathbf{1}}(t), \end{aligned} \tag{3.19}$$

where we have used the equation of motion of $\bar{\mathbf{1}}$ from Eq. (3.7). Then, using the semiclassical definition of an ensemble average in Eq. (3.14), the sum of the singlet and triplet yield is

$$\begin{aligned} \Phi_S + \Phi_T &= \prod_{i=1}^{2} \frac{2}{4\pi} \int d\Omega_{S_i} \prod_{k=1}^{N_i} \frac{2I_{ik}+1}{4\pi} \int d\Omega_{I_{ik}}\, \bar{\rho}(0) \int_0^\infty \left(-\frac{\mathrm{d}}{\mathrm{d}t}\bar{\mathbf{1}}(t)\right) dt \\ &= \prod_{i=1}^{2} \frac{2}{4\pi} \int d\Omega_{S_i} \prod_{k=1}^{N_i} \frac{2I_{ik}+1}{4\pi} \int d\Omega_{I_{ik}}\, \bar{\rho}(0), \end{aligned} \tag{3.20}$$

since when the radical pair recombines, $\bar{k} > 0$ and $\bar{\mathbf{1}}(\infty) = 0$. $\bar{\rho}(0)$ can always be written as normalised linear combination of the classical variables corresponding to the projection operators,

$$\begin{aligned} \bar{P}_S &= \frac{1}{4}\bar{\mathbf{1}} - \mathrm{tr}[\mathbf{T}_{12}], \\ \bar{P}_{T_+} &= \frac{1}{4}\bar{\mathbf{1}} + \frac{1}{2}(S_{1z} + S_{2z}) + S_{1z}S_{2z}, \\ \bar{P}_{T_0} &= \frac{1}{4}\bar{\mathbf{1}} + \mathrm{tr}[\mathbf{T}_{12}] - 2S_{1z}S_{2z}, \\ \bar{P}_{T_-} &= \frac{1}{4}\bar{\mathbf{1}} - \frac{1}{2}(S_{1z} + S_{2z}) + S_{1z}S_{2z}, \end{aligned} \tag{3.21}$$

and so can be expressed as

$$\bar{\rho}(0) = \frac{1}{4Z}\bar{\mathbf{1}}(0) + f(\mathbf{S}_1(0), \mathbf{S}_2(0), \mathbf{T}_{12}(0)). \tag{3.22}$$

Since $f(\mathbf{S}_1(0), \mathbf{S}_2(0), \mathbf{T}_{12}(0))$ is a linear function of the one- and two-electron classical variables, it vanishes when integrated over all possible orientations of the electron spin vectors. Therefore,

$$\Phi_S + \Phi_T = \prod_{i=1}^{2} \frac{2}{4\pi} \int d\Omega_{S_i} \prod_{k=1}^{N_i} \frac{2I_{ik}+1}{4\pi} \int d\Omega_{I_{ik}} \left(\frac{1}{4Z} \bar{\mathbf{1}}(0) + f(\mathbf{S}_1, \mathbf{S}_2, \mathbf{T}_{12}) \right).$$
$$= \frac{1}{4Z} \prod_{i=1}^{2} \frac{2}{4\pi} \int d\Omega_{S_i} \prod_{k=1}^{N_i} \frac{2I_{ik}+1}{4\pi} \int d\Omega_{I_{ik}} = 1, \tag{3.23}$$

and our semiclassical expressions in Eqs. (3.7) and (3.14) are consistent with this exact sum rule.

Returning to the inconsistency in the semiclassical theory highlighted earlier, it is now clear that the most straightforward way to remove the unphysical behaviour will be to modify the equations of motion of the nuclear spins, Eq. (3.3). One way these equations can be changed to recover a physically reasonable model is to replace them with

$$\frac{d}{dt}\mathbf{I}_{ik} = a_{ik} \frac{\sqrt{S_i(S_i+1)}}{|\mathbf{S}_i|} \mathbf{S}_i \times \mathbf{I}_{ik}. \tag{3.24}$$

Now the nuclear spins will always precess around an electron spin vector of length $\sqrt{S_i(S_i+1)}$, and the magnitude of the nuclear spin vector no longer changes, so the electron spin precession will not slow down either. While this correction is admittedly *ad hoc*, it is both physically motivated and necessary to recover the correct behaviour in the limiting case of symmetric recombination.

In this section we have outlined our semiclassical theory of the spin dynamics of radical pairs. We shall now show how it is related to the earlier semiclassical theory of Schulten and Wolynes, before comparing the performance of both semiclassical methods to fully quantum mechanical simulations in Sect. 3.3.

3.2 Schulten–Wolynes Theory

In 1978, Schulten and Wolynes introduced their theory of the spin dynamics of radical pair reactions [1]. It is similar to our theory, outlined in Sect. 3.1, in that it makes the semiclassical approximation of replacing vector operators with classical vectors. However, Schulten and Wolynes made some additional assumptions which further simplify the problem, and in the case of symmetric recombination allow the derivation of analytical expressions for the ensemble averages of some observables.

Firstly, Schulten and Wolynes assumed that there are sufficiently many nuclear spins present that the hyperfine-weighted resultant of all of the nuclear spins on each radical

$$\mathbf{I}_i = \sum_{k=1}^{N_i} a_{ik}\mathbf{I}_{ik}, \tag{3.25}$$

and hence also

$$\bar{\omega}_i = \omega_i + \mathbf{I}_i, \tag{3.26}$$

can be considered constant. This approximation is reasonable at short times, since the electron spin will precess much more quickly than $\mathbf{I}_i$, and becomes exact in the limit that $N_i \to \infty$ [4, 5]. It is equivalent to replacing Eq. (3.24), the equation of motion of the nuclear spins in our semiclassical theory, with

$$\frac{d}{dt}\mathbf{I}_{ik} = 0. \tag{3.27}$$

However, this approximation is not consistent with Newton's third law – the torque exerted on the electron spin by the nuclear spins is not matched by an equal and opposite torque on the nuclear spins by the electron spin.

Secondly, Schulten and Wolynes treated this resultant vector $\mathbf{I}_i$ as a random flight polymer, and assumed a distribution of end-to-end distances $I_i = |\mathbf{I}_i|$ given by [1, 6]

$$g(\mathbf{I}_i) = \left(\frac{\tau_i^2}{4\pi}\right)^{3/2} e^{-I_i^2\tau_i^2/4}, \tag{3.28}$$

where

$$\tau_i^{-2} = \frac{1}{6}\sum_{k=1}^{N_i} a_{ik}^2 I_{ik}(I_{ik}+1) \tag{3.29}$$

This assumption is valid in the same limit as the first, that of a large number of nuclear spins on radical i, due to the central limit theorem. These additional assumptions lead to new expressions for the ensemble average of an observable of a radical pair,

$$\mathrm{A}(t) \simeq \prod_{i=1}^{2} \frac{2}{4\pi}\int d\Omega_{S_i} \int d\mathbf{I}_1\, g(\mathbf{I}_1) \int d\mathbf{I}_2\, g(\mathbf{I}_2)\bar{\rho}(0)\bar{A}(\mathbf{I}_1, \mathbf{I}_2, t), \tag{3.30}$$

and the spin correlation tensor of a singlet-born radical pair,

$$R_{\alpha\beta}^{(i)}(t) \simeq \frac{2}{4\pi}\int d\Omega_{S_i} \int d\mathbf{I}_i\, g(\mathbf{I}_i) S_{i\alpha}(0) S_{i\beta}(\mathbf{I}_i, t). \tag{3.31}$$

The classical variables $\bar{A}(\mathbf{I}_1, \mathbf{I}_2, t)$ and $S_{i\beta}(\mathbf{I}_i, t)$ in Eqs. (3.30) and (3.31) are obtained from the equations of motion in Eqs. (3.4)–(3.7), with $\bar{\boldsymbol{\omega}}_i$ related to $\mathbf{I}_i$ by Eq. (3.26).

When the recombination rates of the radical pair are symmetric and can be factored out of the spin dynamics, the equations of motion of the electron spin vectors can be solved analytically [2]:

$$\mathbf{S}_i(t) = \frac{1-\cos\bar{\omega}_i t}{\bar{\omega}_i^2}\bar{\boldsymbol{\omega}}_i\,(\bar{\boldsymbol{\omega}}_i \cdot \mathbf{S}_i(0)) + \mathbf{S}_i(0)\cos\bar{\omega}_i t + \frac{\sin\bar{\omega}_i t}{\bar{\omega}_i}\bar{\boldsymbol{\omega}}_i \times \mathbf{S}_i(0), \tag{3.32}$$

since $\mathbf{I}_1$ and $\mathbf{I}_2$ and therefore $\bar{\boldsymbol{\omega}}_1$ and $\bar{\boldsymbol{\omega}}_2$ are independent of time. Here $\bar{\omega}_i = |\bar{\boldsymbol{\omega}}_i|$. Using this result, we may use SW theory to find analytical expressions for the ensemble average of any observable which can be written as a function of the electron spin variables. For example, the singlet probability of a singlet-born radical pair may be written as

$$\mathrm{P_S}(t) \simeq \left(\frac{2}{4\pi}\right)^2 \int d\Omega_{S_1} \int d\Omega_{S_2} \int d\bar{\omega}_1\, g(\bar{\omega}_1) \int d\bar{\omega}_2\, g(\bar{\omega}_2) \bar{P}_\mathrm{S}(0) \bar{P}_\mathrm{S}(\bar{\omega}_1, \bar{\omega}_2, t), \tag{3.33}$$

expressed now in terms of $\bar{\omega}_i$ rather than $\mathbf{I}_i$. Here

$$g(\bar{\boldsymbol{\omega}}_i) = \left(\frac{\tau_i^2}{4\pi}\right)^{3/2} e^{-(\bar{\omega}_i^2+\omega_i^2-2\bar{\omega}_i\omega_i\cos\theta_i)\tau_i^2/4}, \tag{3.34}$$

where θ_i is the polar angle of the resultant vector $\bar{\omega}_i$ relative to a static applied field $\boldsymbol{\omega}_1 = \boldsymbol{\omega}_2 = \boldsymbol{\omega}$ of magnitude ω, and

$$\bar{P}_\mathrm{S}(t) = \frac{1}{4}\bar{\mathbf{1}} - \mathrm{tr}[\mathbf{T}_{12}(t)] = \frac{1}{4}\bar{\mathbf{1}} - \mathbf{S}_1(t)\cdot\mathbf{S}_2(t). \tag{3.35}$$

Substituting Eq. (3.32) into Eq. (3.33) and integrating over the initial orientations of the electron spin vectors gives

$$\mathrm{P_S}(t) = \int d\bar{\omega}_1 \int d\bar{\omega}_2\, g(\bar{\omega}_1) g(\bar{\omega}_2) \frac{1}{16}\Bigg[1 + \cos\bar{\omega}_1 t + \cos\bar{\omega}_2 t + \cos\bar{\omega}_1 t \cos\bar{\omega}_2 t + \frac{2}{\bar{\omega}_1\bar{\omega}_2}(\bar{\omega}_1\cdot\bar{\omega}_2)\sin\bar{\omega}_1 t \sin\bar{\omega}_2 t + \frac{1}{\bar{\omega}_1^2\bar{\omega}_2^2}(\bar{\omega}_1\cdot\bar{\omega}_2)^2(1-\cos\bar{\omega}_1 t)(1-\cos\bar{\omega}_2 t)\Bigg]. \tag{3.36}$$

Finally, Rodgers has shown that the integrals over $\bar{\omega}_1$ and $\bar{\omega}_2$ may also be solved analytically [7]. This leads to an extremely lengthy expression, included in Appendix D. A more manageable expression is obtained in the limit of no nuclear spins on one of the radicals, such that $\bar{\boldsymbol{\omega}}_2 = \boldsymbol{\omega}_2 = \boldsymbol{\omega}$ and $g(\bar{\omega}_2) = 1$. In that case [7],

$$\begin{aligned}\mathrm{P_S}(t) &= \frac{\left(1+e^{-t^2/\tau_1^2}\right)}{2\tau_1^2\omega^2}\left(\tau_1^2\omega^2 + 2\cos\omega t - 2\right) + \frac{i\sqrt{\pi}e^{-\tau_1^2\omega^2/4}}{2\tau_1^3\omega^3}(\cos\omega t - 1) f(t,\tau_1,\omega),\\ f(t,\tau_1,\omega) &= \left[\mathrm{erf}\left(\frac{t}{\tau_1} - \frac{i\tau_1\omega}{2}\right) - \mathrm{erf}\left(\frac{t}{\tau_1} + \frac{i\tau_1\omega}{2}\right) + 2\mathrm{erf}\left(\frac{i\tau_1\omega}{2}\right)\right],\end{aligned} \tag{3.37}$$

where $\mathrm{erf}(x)$ is the error function,

$$\mathrm{erf}(x) = \frac{2}{\sqrt{\pi}}\int_0^x e^{-y^2}\mathrm{d}y. \tag{3.38}$$

Similar closed form expressions may be obtained for the components of the spin correlation tensor [2, 7], which are also listed in Appendix D.

When the recombination rates of the radical pair are not equal, it is no longer possible to find analytical expressions for the ensemble averages of observables within SW theory. Nevertheless, it is inexpensive to use Eq. (3.30) to find ensemble averages numerically, as it only requires the solution of at most the 16 coupled differential equations in Eqs. (3.4)–(3.7) since the nuclear spins no longer evolve in time. The singlet yield of a radical pair is even more straightforward to evaluate. Using the quantum mechanical definition of $P_S(t)$ from Eq. (2.38), and introducing the Liouvillian superoperator

$$-\hat{\hat{L}}\hat{\rho} = -i[\hat{H}, \hat{\rho}] - \{\hat{K}, \hat{\rho}\}, \tag{3.39}$$

where $[\cdots]$ indicates a commutator and $\{\cdots\}$ an anticommutator, the singlet yield may be written as

$$\Phi_S = k_S \int_0^\infty \mathrm{tr}[\hat{P}_S e^{-\hat{\hat{L}}t} \hat{\rho}(0)]\, \mathrm{d}t. \tag{3.40}$$

The integral over time can be easily evaluated:

$$\Phi_S = k_S\, \mathrm{tr}[\hat{P}_S \hat{\hat{L}}^{-1} \hat{\rho}(0)]. \tag{3.41}$$

Then, applying Schulten and Wolynes' approximations, the quantum mechanical trace over the nuclear spin states can be replaced by integrals over the resultant vectors $\mathbf{I}_1$ and $\mathbf{I}_2$, leaving

$$\Phi_S = \int d\mathbf{I}_1\, g(\mathbf{I}_1) \int d\mathbf{I}_2\, g(\mathbf{I}_2)\, k_S\, \mathrm{tr}_\mathbf{S}[\hat{P}_S \hat{\hat{L}}(\mathbf{I}_1, \mathbf{I}_2)^{-1} \hat{\rho}(0)], \tag{3.42}$$

where $\mathrm{tr}_\mathbf{S}$ indicates a trace over the four dimensional Hilbert space of the two electron spins. This expression allows the straightforward and efficient calculation of the singlet yield of a radical pair, and is used in Chap. 6 to analyse the magnetic field effect on the electroluminescence of organic light emitting diodes.

3.3 Comparison of Methods

3.3.1 A Simple Radical Pair

In order to compare the three methods we have now described it is helpful to apply them to a range of model systems. The first example is taken from Ref. [2], where the authors used a simple radical pair with no recombination or electron spin coupling and nuclear spins on only one of the two radicals to establish the accuracy of the two semiclassical methods against exact quantum mechanical results. Every nuclear spin had the same quantum number, $I = 1/2$, and a hyperfine coupling constant taken from

Table 3.1 The hyperfine coupling constants used in the model radical pair in Ref. [2]

k	$a_{1k}/\lvert\gamma_e\rvert$ (mT)	k	$a_{1k}/\lvert\gamma_e\rvert$ (mT)
1	−0.999985	9	0.358593
2	−0.736925	10	0.869386
3	0.51121	11	−0.232996
4	−0.0826998	12	0.0388327
5	0.0655341	13	0.661931
6	−0.562082	14	−0.930856
7	−0.905911	15	−0.893077
8	0.357729	16	0.0594001

a uniform distribution between $-1 \leq a_{1k}/|\gamma_e| \leq 1$ mT. These constants are listed in Table 3.1. Three components of the spin correlation tensor of a radical with $N_1 = 1$, 4, and 16 of these nuclear spins in a magnetic field of strength $B = 0.5$ mT applied along the z axis were calculated quantum mechanically (QM), and using both the semiclassical (SC) theory and Shulten-Wolynes (SW) theory. The results are shown in Fig. 3.3, along with the singlet probability of the corresponding radical pair when $N_2 = 0$.

When $N_1 = 1$, coherent oscillations are observed in the QM calculation of the components of the spin correlation tensor and the singlet probability, which unsurprisingly are not captured by either semiclassical approach. However, the SC theory does track the average of these oscillations reasonably closely. We can hardly expect SW theory to be accurate in this case, since it is derived using approximations that only become exact when $N_1 \to \infty$. The coherent oscillations are damped as the number of nuclear spins increases, and by the time $N_1 = 16$ only the semiclassical precession remains. This is reproduced quantitatively by the SC theory; however, SW theory is only accurate at short times $t < 2\tau_1$, before significant precession of the nuclear spins has occurred [2].

The SC method is also far less computationally expensive than the full QM calculation by the time $N_1 = 16$, since the time required increases only linearly with the number of nuclear spins, rather than exponentially. The relative CPU time taken for the SC and QM calculations as N_1 increases is shown in Fig. 3.4.[3] For small numbers of nuclear spins, $N_1 < 10$, the SC method is less efficient due to the number of samples required to converge the integrals in Eq. (3.14). However, as N_1 increases the SC calculation becomes both quantitatively accurate and more efficient than the QM calculation. The SW result, of course, requires no computation at all, as we have found analytical expressions for the singlet probability in Eq. (3.37) and the components of the spin correlation tensor in Appendix D. Therefore, there exists a range

[3] It should be noted that these QM calculations were performed deterministically, rather than with the more efficient stochastic method outlined in Sect. 2.5. While using the latter method would probably mean that the cross-over in efficiency occurred for a slightly larger value of N_1, the SC must necessarily become more efficient at some point.

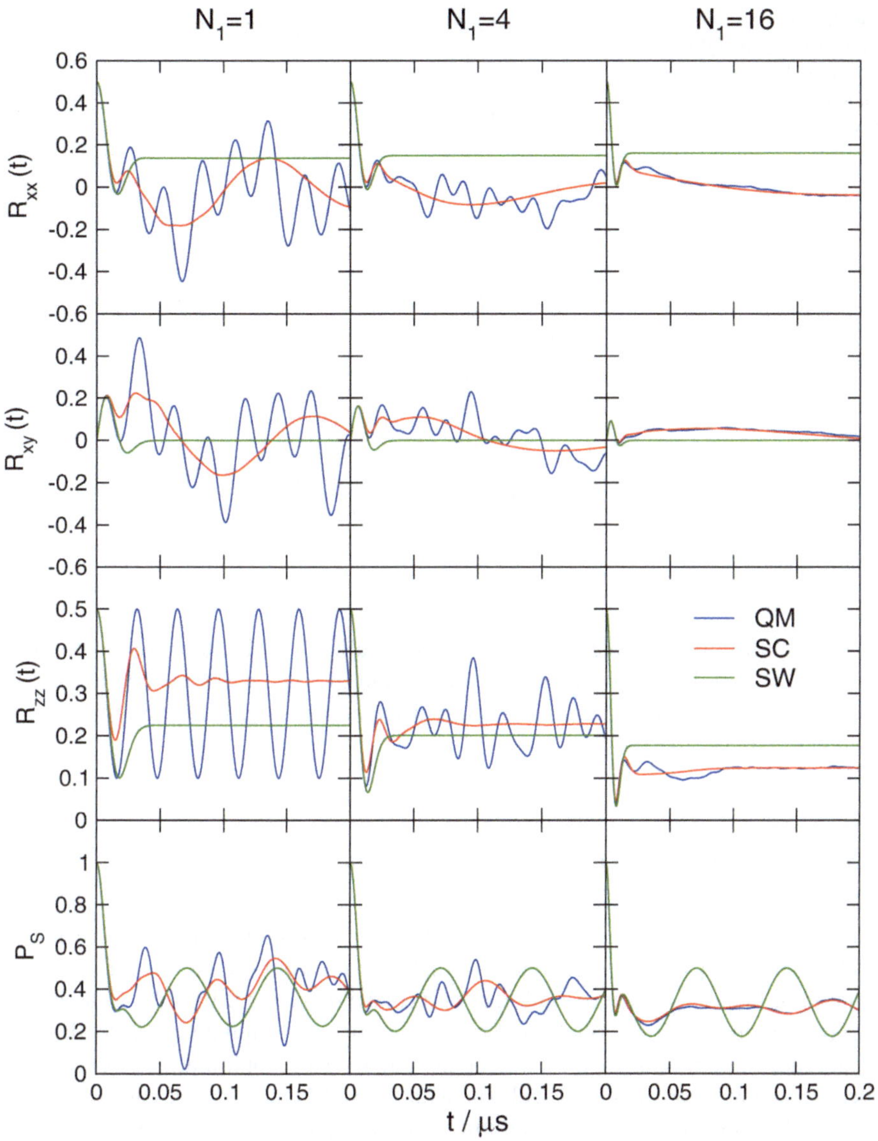

Fig. 3.3 The xx, xy, and zz components of the spin correlation tensor of a radical with $N_1 = 1, 4$, and 16 nuclear spins, and the singlet probability of the corresponding radical pair with $N_2 = 0$, no recombination, and no electron spin coupling. They hyperfine coupling constants for radical 1 are given in Table 3.1. Adapted with permission from Ref. [2]

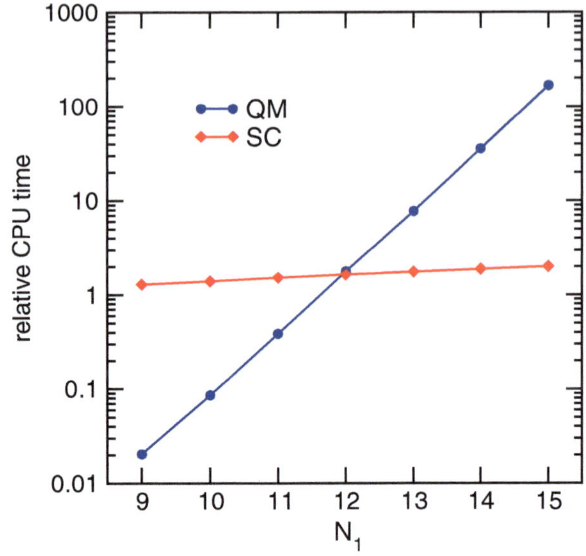

Fig. 3.4 The relative CPU time taken by the QM and SC methods as the number of nuclear spins in radical 1 increases. Adapted with permission from Ref. [2]

of applicability for each method: when N_1 is small, the QM calculation is possible; when N_1 is very large SW theory is reliable and in some cases analytical; and for intermediate N_1 the SC theory is far more efficient than a QM calculation and more accurate than SW theory.

3.3.2 Asymmetric Recombination Rates

For the model radical pair in Sect. 3.3.1, the modified equation of motion for the nuclear spins, Eq. (3.24), is identical to that derived from the Heisenberg equation of motion, Eq. (3.3), since recombination has been neglected. In order to assess the accuracy of our modified equation of motion, we need to simulate a slightly different model radical pair, with asymmetric singlet and triplet recombination rates. For this purpose, we will consider a radical pair with $N_1 = 12$ and $N_2 = 0$, with $I = 1/2$ for each nuclear spin and hyperfine coupling constants taken from Table 3.1, again with no electron spin coupling.

The QM, SC, and SW singlet probability of this radical pair are shown in Fig. 3.5 for three different values of recombination asymmetry $\Delta k = 1/4(k_\mathrm{S} - k_\mathrm{T})$, and at two different applied magnetic field strengths, $B = 0.5$ and 4 mT. In each case, the average recombination rate $\bar{k}/|\gamma_e| = 1/4(k_\mathrm{S} + 3k_\mathrm{T})/|\gamma_e| = 0.1$ mT, as we move from symmetric rates, $\Delta k = 0$, through an intermediate case, $\Delta k = \bar{k}/2$, to fully asymmetric recombination, $\Delta k = \bar{k}$. In every scenario, there is quantitative agreement between the SC and QM results; certainly the SC theory seems no less accurate when $\Delta k \neq 0$. By contrast, the SW results show a pronounced Zeeman oscillation

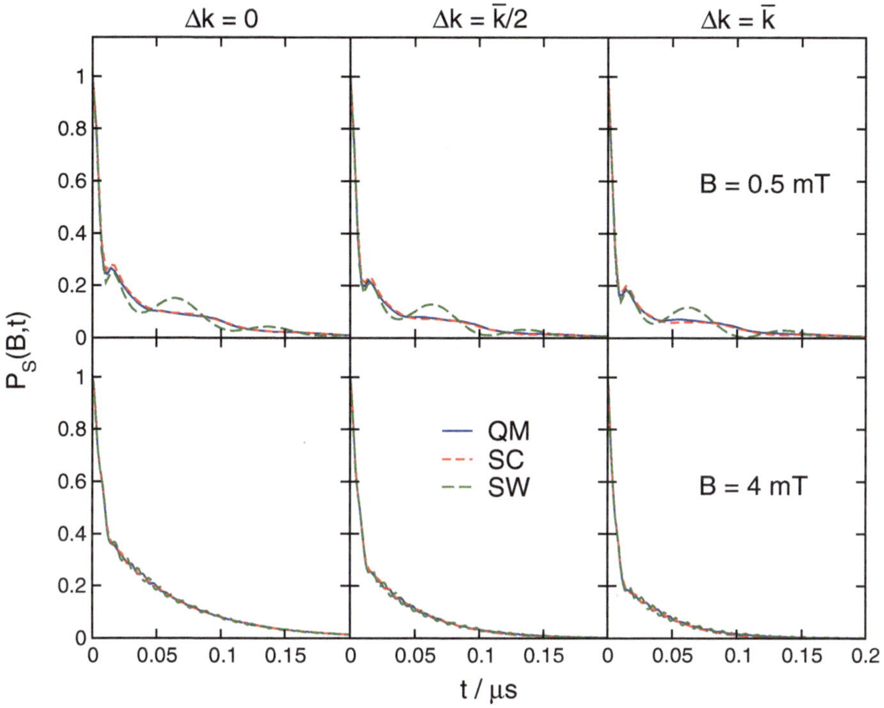

Fig. 3.5 The quantum mechanical, semiclassical, and Schulten–Wolynes singlet probabilities for a model radical pair with $N_1 = 12$ and hyperfine coupling constants taken from Table 3.1, for three values of Δk and two different magnetic field strengths. $\bar{k}/|\gamma_e| = 0.1$ mT in every case. Data is taken from Ref. [3]

with period $2\pi/|\gamma_e|B$ μs, which do not appear in either the SC or QM simulations. This is also apparent in the bottom right panel of Fig. 3.3. It appears that including the nuclear spin dynamics in the SC theory damps these oscillations and so produces results which are much closer to the correct QM singlet probability.

It is also instructive to compare the exact singlet yield of the radical pair to that calculated semiclassically over a range of magnetic field strengths, as in Fig. 3.6. In the case of both symmetric and asymmetric recombination, the SC theory performs much better in the low field region than SW theory, which significantly overestimates the singlet yield in the absence of a magnetic field. In the case of symmetric recombination, both semiclassical theories show excellent agreement with the quantum mechanical results at high fields. As can be seen from Fig. 3.5, the frequency of the spurious Zeeman oscillation in the SW results increases with B, and so the absence of nuclear spin dynamics to damp these oscillations becomes less important at high fields.

Strangely, when the recombination rates are not equal the SC method shows slightly poorer agreement with the exact results than SW theory in the high field region. However, this discrepancy is much smaller than the error in the SW results at

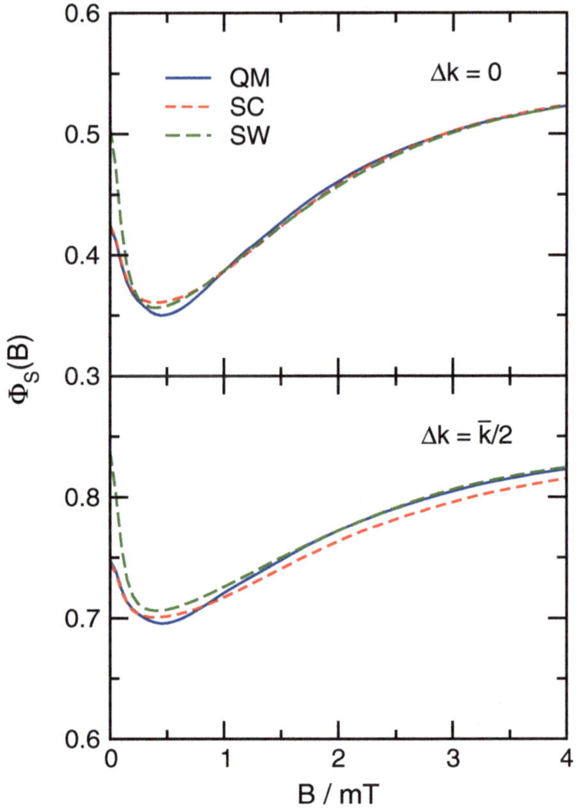

Fig. 3.6 The quantum mechanical, semiclassical, and Schulten–Wolynes singlet yield of the same radical pair as Fig. 3.5 for two values of Δk. When $\Delta k = \bar{k}$, the singlet yield equals one for all B. Data is taken from Ref. [3]

low fields. Furthermore, it should be re-emphasised that we expect the accuracy of the SC theory to improve as the number of nuclear spins increases. Overall, it appears that the modification to the equations of motion for the nuclear spins required when the recombination rates are not symmetric does not significantly reduce the accuracy of the SC theory, and it remains the best method to use for intermediate numbers of nuclear spins.

3.3.3 *Exchange Coupling*

Finally, Fig. 3.7 compares the quantum mechanical results with those found using the semiclassical theory and Schulten–Wolynes theory for the model problem introduced in Sect. 2.5, which includes exchange coupling between the two electron spins. Unlike the previous models, here neither semiclassical method is quantitatively accurate. Of the two, however, the SC theory is qualitatively better than SW theory, correctly predicting the position of the minimum in the singlet yield, and in much closer

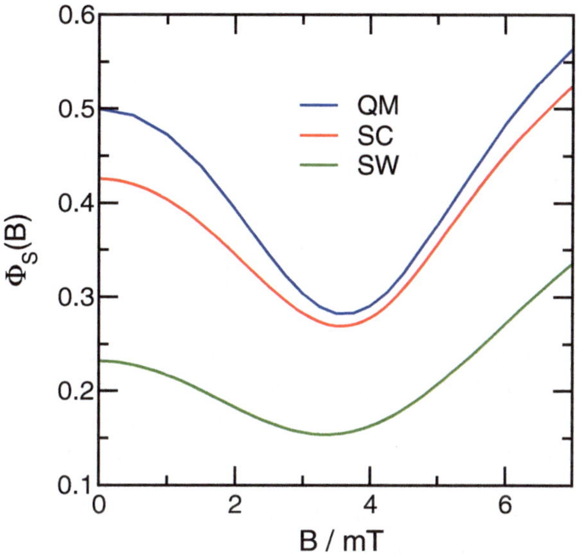

Fig. 3.7 The relative singlet yield of the model radical pair described in Sect. 2.5 as a function of magnetic field strength B, calculated using the stochastic quantum mechanical approach (QM) introduced in Sect. 2.5, the semiclassical method (SC) presented in Sect. 3.1, and a variation of Schulten–Wolynes theory (SW) outlined in Sect. 3.2. Data is taken from Ref. [8]

agreement with the QM results across the entire range of magnetic field strengths. Clearly in this case, where the hyperfine interactions, exchange coupling and the applied magnetic field are all of a similar magnitude, semiclassical models are not sufficient to accurately capture the fine details of the spin dynamics. The quantum mechanical method is required for applications where all of these interactions are important, such as the one described in Chap. 4.

3.4 Relaxation

At the end of Chap. 2 we attempted to introduce relaxation into our quantum mechanical simulations. There we found that the processes which give rise to relaxation are extremely difficult to describe accurately, and even simple phenomenological models of relaxation greatly increase the computational time required by simulations. While a detailed account of relaxation remains difficult in the semiclassical theory, a phenomenological one analogous to the Bloch equations introduced in Sect. 2.7.1 is straightforward to include and does not increase the numerical complexity of the problem.

The semiclassical equivalent to the Bloch equations simply consists of adding the following terms to Eqs. (3.4)–(3.6):

$$\frac{d}{dt}\mathbf{S}_1 = \cdots - \mathbf{R}^{(1)}\mathbf{S}_1, \tag{3.43}$$

$$\frac{d}{dt}\mathbf{S}_2 = \cdots - \mathbf{R}^{(2)}\mathbf{S}_2, \tag{3.44}$$

$$\frac{d}{dt}\mathbf{T}_{12} = \cdots - \mathbf{R}^{(1)}\mathbf{T}_{12} - \mathbf{T}_{12}\mathbf{R}^{(2)}. \tag{3.45}$$

Here $\cdots$ indicates the evolution and recombination terms also present in the equations of motion, and

$$\mathbf{R}^{(i)} = \begin{pmatrix} 1/T_2^{(i)} & 0 & 0 \\ 0 & 1/T_2^{(i)} & 0 \\ 0 & 0 & 1/T_1^{(i)} \end{pmatrix}. \tag{3.46}$$

The electron spins are assumed to relax independently of one another, with the x and y components of electron spin i relaxing at a rate $1/T_2^{(i)}$ and the z component at a rate $1/T_1^{(i)}$, as is conventional. We have assumed that the magnetic field is sufficiently small that there is no polarisation of the equilibrium distribution of either electron spin; that is, we assume that $\langle \hat{S}_{eq} \rangle = 0$. This assumption is reasonable for every application of the semiclassical theory presented in this thesis.

These equations produce physically reasonable results: the two-electron spin variables relax at the sum of the rates of the corresponding single spin variables; relaxation reduces the magnitude of $\mathrm{tr}[\mathbf{T}_{12}]$ and so, when the radical pairs do not recombine, causes $\bar{P}_S$ and $\bar{P}_T$ to tend to their equilibrium values of 1/4 and 3/4 respectively; and the equation of motion of $\bar{\mathbf{1}}$ is unchanged when relaxation is included. The equations can be further simplified in the case of very small applied magnetic fields, where we may make the approximation that $T_1^{(i)} \simeq T_2^{(i)}$. The relaxation of the nuclear spins has been neglected on the grounds of their far smaller gyromagnetic ratios.

While this method of treating relaxation is admittedly crude, the fact that it is possible with essentially no extra computational cost nonetheless represents a considerable advantage of the semiclassical theory over quantum mechanical calculations. Furthermore, we will demonstrate in Sect. 5.2 that this approach is sufficient to reproduce the experimentally measured Earth-strength magnetic field effect on the survival probability of a carotenoid-porphyrin-fullerene triad in solution. More explicit approaches to accounting for electron spin relaxation in the semiclassical theory have been considered recently and show some promise [9], but do not yet represent a significant improvement upon the simple theory outlined here.

3.5 Conclusion

In this chapter, we have introduced the semiclassical theory as a practical solution to the exponential scaling problem of quantum mechanical calculations. We have demonstrated numerically that the semiclassical theory is exact in the limit of constant magnetic fields, and is therefore also correct when $N \to \infty$ or $B \to \infty$. We have also shown how our theory reduces to Schulten and Wolynes' theory with two additional approximations, and how SW theory may be formulated for a radical pair with asymmetric recombination rates. We compared the performance of the two semiclassical methods to exact quantum mechanical simulations for three different model problems, showing the SC theory to be an improvement on SW theory in all cases, and in quantitative agreement with QM results in the absence of electron spin coupling. Finally, we discussed how relaxation of the electron spins could be accounted for in the SC theory.

We are now in a position to apply these three methods to a range of different problems in the next three chapters. In Chap. 4, we will use the efficient quantum mechanical method described in Sect. 2.5 to analyse the the mechanisms involved in charge recombination along molecular wires. In Chap. 5 the semiclassical theory introduced in Sect. 3.1 is shown to reproduce experimental measurements of an Earth-strength magnetic field effect, and is used to assess the suitability of a cryptochrome-based radical pair to act as a biological compass. Finally, in Chap. 6 we shall identify the physical interactions which govern the magnetoelectroluminscence of organic light emitting diodes using the formulation of SW theory presented at the end of Sect. 3.2.

References

1. Schulten, K., & Wolynes, P. G. (1978). Semi-classical description of electron-spin motion in radicals including effect of electron hopping. *Journal of Chemical Physics*, *68*, 3292–3297.
2. Manolopoulos, D. E., & Hore, P. J. (2013). An improved semiclassical theory of radical pair recombination reactions. *Journal of Chemical Physics*, *139*, 124106.
3. Lewis, A. M., Manolopoulos, D. E., & Hore, P. J. (2014). Asymmetric recombination and electron spin relaxation in the semiclassical theory of radical pair reactions. *Journal of Chemical Physics*, *141*, 044111.
4. Chen, G., Bergman, D. L., & Balents, L. (2007). Semiclassical dynamics and long-time asymptotics of the central-spin problem in a quantum dot. *Physical Review B*, *76*, 045312.
5. Erlingsson, S. I., & Nazarov, Y. V. (2004). Evolution of localized electron spin in a nuclear spin environment. *Physical Review B*, *70*, 205327.
6. Flory, P. J. (1969). *Statistical mechanics of chain molecules*. Geneva: Interscience Publishers.
7. Rodgers, C. T. (2007). *magnetic field effects in chemical systems*. PhD thesis, University of Oxford, Oxford.
8. Lewis, A. M., Fay, T. P., & Manolopoulos, D. E. (2016). An effcient quantum mechanical method for radical pair recombination reactions. *Journal of Chemical Physics*, *145*, 244101.
9. Vanderstichele, B. (2017). *Electron spin relaxation in the semiclassical theory of radical pair reactions*. Part II thesis, University of Oxford, Oxford.

Chapter 4
Molecular Wires

In recent years, molecular wires have been the subject of significant interest and investigation [1, 2]. A common class of molecular wires are those with a D–B–A structure – an electron donor separated from an electron acceptor by a molecular "bridge", typically an oligomer, which allows precise control over the separation between the electron donor and acceptor [3]. Molecular wires undergo charge separation following photoexcitation to form a radical pair, $D^{\bullet+}$–B–$A^{\bullet-}$, mimicking the efficient long range charge transport found in the photosynthetic reaction centre [4, 5]. There are a number of possible applications for molecular wires; in particular, efficient 'wire-like' charge separation to form a long-lived radical pair is a highly desirable feature in chemical solar energy conversion systems [3, 6]. These systems depend on the rapid transport of electrons from the chromophores where light is absorbed to a reaction centre, where they must survive for long enough to catalyse a reaction, such as the reduction of CO_2 [6].

In order to design molecular wires which are suitable for this purpose, an understanding of the mechanisms by which charge recombination along them occurs is crucial [7]. However, this is a complex problem: the distance dependence of the rate of electron transfer has been shown to vary with a number of factors, including the geometry of the molecular wire as well as its energetics [1, 8, 9]. Usually, the rate of electron transfer is found to decrease exponentially as the distance between the donor and acceptor sites increases [10, 11], although the characteristic length of this dependence can vary significantly [12, 13]. However, there are a number of examples where this exponential dependence on distance is not observed, with the rate of electron transfer depending only very weakly on the separation of the donor and acceptor [14, 15].

A. Lewis, *Spin Dynamics in Radical Pairs*, Springer Theses,
https://doi.org/10.1007/978-3-030-00686-0_4

In this chapter, we shall begin by examining the mechanisms which give rise to these two different types of behaviour, superexchange and incoherent hopping.[1] We will then turn our attention to a particular set of molecular wires, $PTZ^{\bullet+}$–Ph_n–$PDI^{\bullet-}$ with $n = 2 - 5$, whose triplet recombination yields and "radical pair yields" – the survival probability of the radical pair 50 ns after the initial photoexcitation laser pulse – have been measured at a range of different applied magnetic field strengths [16]. By fitting quantum mechanical calculations of the triplet and radical pair yields of these radical pairs to the experimental measurements, it is possible to extract their singlet and triplet recombination rate constants. In order to do this, we must determine the parameters which define the spin Hamiltonian of these radical pairs and consider the experimental evidence which constrains their recombination rate constants. Finally we shall use the rate constants obtained from these simulations to shed light on the mechanisms of radical pair recombination which operate along these molecular wires.

4.1 Mechanisms of Charge Recombination

There are two limiting mechanisms of charge recombination which operate in molecular wires: the superexchange mechanism and the incoherent hopping mechanism [8, 17]. In the superexchange mechanism, recombination occurs by an electron tunnelling from the $A^{\bullet-}$ radical to the $D^{\bullet+}$ radical in a single step via the superexchange coupling between the electron and hole, which is mediated by orbitals on the bridge [18]. The magnitude of this coupling, and therefore the rate of electron transfer by this mechanism, is expected to decrease exponentially with increasing donor-acceptor separation [19, 20]. By contrast, the incoherent hopping mechanism is a two step process, in which an electron hops from the bridge to the $D^{\bullet+}$ radical, followed by a second electron hopping from the $A^{\bullet-}$ radical onto the bridge. The rate of charge transfer by this mechanism is governed by the energy gap between the initial charge separated state of the molecular wire, $D^{\bullet+}$–B–$A^{\bullet-}$, and the intermediate formed by the first electron hopping step, D–$B^{\bullet+}$–$A^{\bullet-}$. Provided that this energy gap is small, the rate of electron transfer is approximately independent of the radical pair separation [9, 21].

The charge separated state of a molecular wire is formed initially in the singlet spin state, $^{S}[D^{\bullet+}$–B–$A^{\bullet-}]$, but then undergoes intersystem crossing to form the triplet state, as shown in Fig. 1.2 [22]. In general, charge recombination along a molecular wire in the singlet state will occur at a different rate, and potentially by a different mechanism, than along a wire in the triplet state. Therefore, the intersystem crossing between these spin states can play a significant role in the overall rate of charge transfer. The interconversion between spin states, and hence the overall charge recombination rate, is influenced by the application of an external magnetic field, because the intersystem crossing is mediated by the hyperfine interactions between

[1] Parts of this chapter have been reproduced with permission from Fay, T. P., Lewis, A. M. & Manolopoulos, D. E. *Journal of Chemical Physics* **147**, 064107 (2017), https://aip.scitation.org/doi/10.1063/1.4997482.

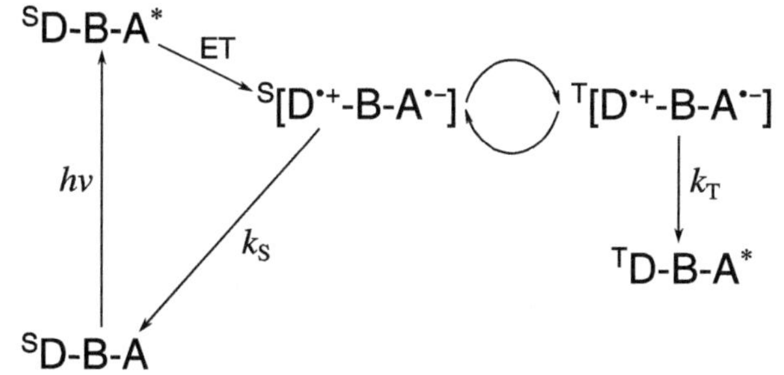

Fig. 4.1 A typical charge recombination reaction along a donor-bridge-acceptor (D–B–A) molecular wire. k_S and k_T are the recombination rate constants for the singlet and triplet states of the radical pair, and the curved arrows represent hyperfine-mediated intersystem crossing between those spin states. The symbol * represents an excited state of the molecule with no unpaired electrons

the electron and nuclear spins on the donor and acceptor. As a result, the magnetic field dependence of experimental observables such as the triplet yield of the recombination reaction can be used to shed light on the rate and mechanism of recombination of the singlet and triplet radical pairs (Fig. 4.1).

In a recent experimental study [16], Weiss et al. used transient absorption spectroscopy to measure the overall recombination rate constants of the $PTZ^{\bullet+}$–Ph_n–$PDI^{\bullet-}$ molecular wires with $n = 1$–5. These consist of a phenothiazine (PTZ) donor, a perylene-3,4:9,10-bis(dicarboximide) (PDI) acceptor, and a bridge of n p-phenylene rings between them. They found evidence which suggested that a change in the mechanism of recombination occurs as the length of the bridge is increased. For wires with short bridges, the rate constants decreased exponentially with increasing bridge length, consistent with the superexchange mechanism. However, the recombination rates of the longer wires increased slightly as the bridge length increased, indicating a change in mechanism to incoherent hopping [16].

The overall recombination rate constants obtained from transient absorption spectroscopy contain contributions from both the singlet and triplet recombination pathways, but their relative importance was not determined by these experiments [16]. This information is crucial to understanding the operation of these wires, since there is no reason to expect that the singlet and triplet states of the radical pair will recombine by the same mechanism. By fitting the experimentally observed magnetic field effects (MFEs) in the triplet and radical pair yields to quantum mechanical simulations, we shall disentangle the contributions of the singlet and triplet pathways to the overall recombination rate, and reveal the likely mechanisms by which these pathways operate.

4.2 Simulation Details

Weiss et al. probed the spin dynamics of the $PTZ^{\bullet+}$–Ph_n–$PDI^{\bullet-}$ molecular wires by measuring the magnetic field effect on the triplet yield and what they term the "radical pair yield" – the survival probability of the radical pair 50 ns after the initial photoexcitation laser pulse. The triplet yield is defined as

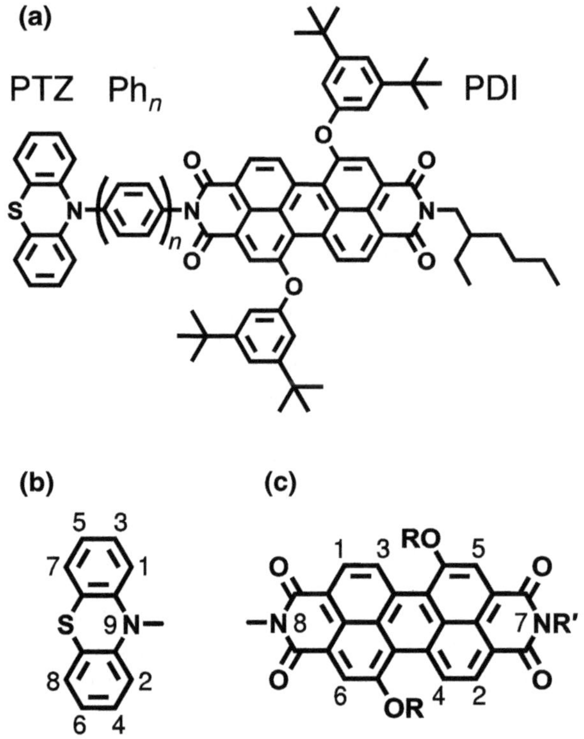

Fig. 4.2 **a** The chemical structure of the $PTZ^{\bullet+}$–Ph_n–$PDI^{\bullet-}$ molecular wires. **b** The positions of the nuclei corresponding the hyperfine coupling constants of the $PTZ^{\bullet+}$ radical listed in Table 4.1. **c** The positions of the nuclei corresponding the hyperfine coupling constants of the $PDI^{\bullet-}$ radical listed in Table 4.2

$$\Phi_{\mathrm{T}} = k_{\mathrm{T}} \int_0^{\infty} \mathrm{P}_{\mathrm{T}}(t)\,\mathrm{d}t, \tag{4.1}$$

where $\mathrm{P_T}(t)$ is the probability of finding the radical pair in the triplet state at time t, and the radical pair yield is simply

$$\Phi_{\mathrm{RP}} = \mathrm{tr}[\hat{\rho}(t)], \tag{4.2}$$

evaluated at $t = 50$ ns, where $\hat{\rho}(t)$ is the density operator of the radical pair. Since there is significant exchange coupling between the unpaired electrons in the $PTZ^{\bullet+}$–Ph_n–$PDI^{\bullet-}$ radical pairs, the semiclassical methods of evaluating these yields described in Chap. 3 are unlikely to be reliable, as demonstrated in Sect. 3.3.3. Therefore, we shall use the stochastic quantum mechanical method outlined in Sect. 2.5 to simulate the triplet and radical pair yields. The latter can be calculated using Eq. (2.58) with $\hat{A} = \hat{1}$, while setting $\hat{A} = \hat{P}_{\mathrm{T}}$ gives the triplet probability $\mathrm{P_T}(t)$ required to calculate the triplet yield.

In order to evaluate Eq. (2.58), we must define the parameters that enter the Hamiltonian in Eq. (2.1) and the recombination operator in Eq. (2.31). The spin evolution of the $PTZ^{\bullet+}$–Ph_n–$PDI^{\bullet-}$ radical pair shown in Fig. 4.2 is governed by the interactions of the electron spins with each other, the applied magnetic field, and the nuclear spins to which they are coupled. We therefore need to know the strength J of the exchange coupling between the two electron spins and the hyperfine coupling constants a_{ik} of the two radicals.

The magnitudes of the hyperfine coupling constants have been measured using electron spin resonance (ESR) spectroscopy for the $PTZ^{\bullet+}$ radical [23], and electron nuclear double resonance (ENDOR) spectroscopy for the $PDI^{\bullet-}$ radical [24]. These measured values are compared with the hyperfine couplings obtained from B3LYP [25, 26] density functional theory (DFT) calculations in Tables 4.1 and 4.2. For the purposes of these calculations, the O-R and N-R′ sidechains in PDI were replaced with O-H and N-H groups.

The agreement between the experimental and calculated coupling constants is not especially good. This is highlighted by comparing the effective hyperfine fields, B_{hyp}. For $PTZ^{\bullet+}$, the experimental hyperfine field is 0.96 mT, while the calculations suggest an effective field of 0.64 mT; for $PDI^{\bullet-}$, the experimental and calculated hyperfine fields are 0.27 and 0.34 mT respectively. The results for the PTZ radical are particularly poor, which may in part be due to the fact the EPR-II basis set normally used to calculate hyperfine constants cannot be applied to this radical, as it

Table 4.1 The hyperfine coupling constants of the $PTZ^{\bullet+}$ radical in mT. Experimental data is taken from Ref. [23]; DFT calculations were performed using the B3LYP functional and cc-PV5Z basis set

k	Nucleus	$\|a_k/\gamma_e\|$ (Expt)	$a_k/\|\gamma_e\|$ (DFT)
1	H	0.113	−0.0753
2	H	0.113	−0.0753
3	H	0.050	−0.0813
4	H	0.050	−0.0813
5	H	0.249	−0.2247
6	H	0.249	−0.2247
7	H	0.050	0.0503
8	H	0.050	0.0503
9	N	0.634	0.3917

Table 4.2 The hyperfine coupling constants of the $PDI^{\bullet-}$ radical in mT. Experimental data is taken from Ref. [24]; DFT calculations were performed using the B3LYP functional and EPR-II basis set

k	Nucleus	$\|a_k/\gamma_e\|$ (Expt)	$a_k/\|\gamma_e\|$ (DFT)
1	H	0.0785	0.0658
2	H	0.0785	0.0658
3	H	0.172	−0.2263
4	H	0.172	−0.2263
5	H	0.0575	0.1351
6	H	0.0575	0.1351
7	N	0.0621	−0.0348
8	N	0.0621	−0.0348

is only parametrised for Period II elements [27]. For this radical we used the larger but hyperfine-unoptimised cc-PV5Z orbital basis set instead [28, 29].

Since the DFT calculations significantly underestimate the size of the hyperfine field on the PTZ radical, we found that it was impossible to quantitatively reproduce the experimental MFEs using the coupling constants obtained from DFT. However, DFT is able to provide the signs of the hyperfine coupling constants, which are not available from the experimental ESR [23] or ENDOR [24] data. Therefore, in our simulations we used hyperfine coupling constants with the magnitude taken from the experimental data and the sign inferred from the DFT calculations.

The exchange coupling constants of these wires can be obtained from the MFEs in the triplet and radical pair yields measured by Weiss et al. [16]. In their experiments, they found a maximum in the triplet yield of the shorter molecular wires ($n = 2, 3$) as the applied magnetic field was increased. This can be understood in terms of the relative energy levels of the singlet and triplet states, as discussed in Sect. 1.3.3. When there is no applied field, or if the applied field is very large, the singlet state is separated in energy from all of the triplet states, limiting singlet-triplet interconversion. However, on resonance, when $B = 2J/|\gamma_e|$, the singlet state in which the radical pair is formed is isoenergetic with the $|T_-\rangle$ state. This results in more efficient intersystem crossing and a maximum in the triplet yield. Similarly, a minimum is observed in the radical pair yield of the longer molecular wires ($n = 4, 5$) at the resonance condition. Therefore, by determining the strength of the applied field at which the triplet (radical pair) yield is largest (smallest), it is possible to infer the magnitude of $2J$. The values of $2J$ for each of the wires were determined in this way in Ref. [16], and are listed in Table 4.3. Note that here we have assumed that the exchange coupling is antiferromagnetic ($J > 0$), informed by ESR experiments on a similar molecular wire [30]. In fact, identical MFEs would be observed if the coupling were antiferromagnetic, with the singlet state becoming degenerate with the $|T_+\rangle$ state, rather than the $|T_-\rangle$ state, at the resonance condition $B = -2J/|\gamma_e|$.

The singlet and triplet recombination rates of the $PTZ^{\bullet+}$–Ph_n–$PDI^{\bullet-}$ molecular wires have not been measured directly, but there are experimental observations which provide constraints on them. Firstly, a minimum in the radical pair yield on resonance, rather than a maximum, implies that the triplet recombination rate is faster than the singlet rate, $k_T > k_S$. Secondly, the radical pair lifetime of each molecular wire has been measured in the absence of a magnetic field by monitoring the decay of the 720 nm absorption band of $PDI^{\bullet-}$ [16]. These are listed in Table 4.3.

For each molecular wire, this lifetime imposes one constraint on the two unknowns k_S and k_T. This is illustrated in Fig. 4.3, which shows the line in the k_S, k_T plane along which our calculations reproduce the experimental radical pair lifetime of the

Table 4.3 Exchange coupling constants and the zero-field recombination lifetimes of the $PTZ^{\bullet+}$–Ph_n–$PDI^{\bullet-}$ molecular wires, both taken from Ref. [16]

n	2	3	4	5
$2J/\lvert\gamma_e\rvert$ mT	170	31	6.4	1.5
τ / ns	21	330	217	121

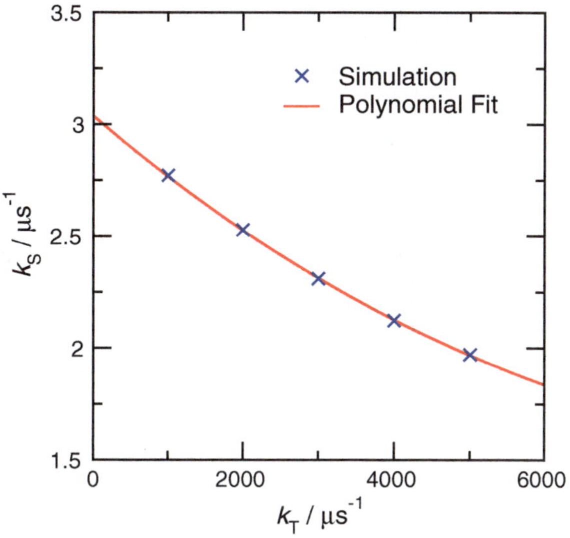

Fig. 4.3 Five pairs of values of (k_S, k_T) which reproduce the overall recombination lifetime of $PTZ^{\bullet +}$–Ph_3–$PDI^{\bullet -}$ in the absence of a magnetic field are shown in blue. The polynomial fit to these points is shown in red, and is defined by $k_S = ak_T^2 + bk_T + c$. The constants a, b, and c for each molecular wire are given in Table 4.4

Table 4.4 The coefficients of the polynomial $k_S = ak_T^2 + bk_T + c$ which defines the (k_S, k_T) parameter space consistent the experimental radical pair lifetime of each molecular wire in the absence of a magnetic field

n	$a/\mu s$	b	$c/\mu s^{-1}$
2	8.540×10^{-11}	-9.375×10^{-6}	47.620
3	1.372×10^{-8}	-2.821×10^{-4}	3.034
4	2.116×10^{-6}	-6.983×10^{-3}	4.634
5	1.194×10^{-3}	-0.215	9.923

$PTZ^{\bullet +}$–Ph_3–$PDI^{\bullet -}$ molecular wire in the absence of a magnetic field. This line may be fit accurately by writing k_S as a quadratic function of k_T. A similar quadratic relationship can be found between k_S and k_T for each of the other molecular wires; the parameters which define these curves are listed in Table 4.4. Since these functions determine k_S for a given k_T, we now only have a single free parameter to vary for each wire to reproduce the MFEs observed by Weiss et al. [16].

4.3 Results

4.3.1 Shorter Wires

For the $n = 2$ and 3 wires, Weiss et al. measured the triplet yield of the radical pair recombination reaction as a function of the applied magnetic field strength, and reported this as the relative triplet yield, $\Phi_T(B)/\Phi_T(0)$. We have simulated these

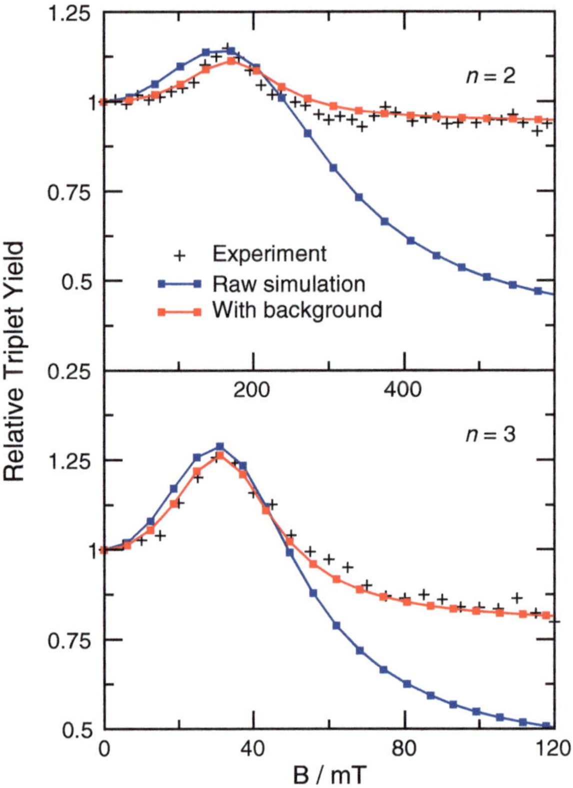

Fig. 4.4 The triplet yield of $PTZ^{\bullet +}$–Ph_n–$PDI^{\bullet -}$ as a function of the strength of the applied magnetic field, relative to the triplet yield in the absence of a field, for $n = 2$ above and $n = 3$ below. The blue curve shows the results of the raw spin dynamics simulations, and the red line the results of the simulations which include a field-independent background contribution to the triplet yield. The experimental data is taken from Ref. [16]

experiments using the parameters defined in Sect. 4.2 in the spin Hamiltonian, taking k_T as an adjustable parameter to fit the experimental data and specifying k_S in terms of k_T in accordance with Table 4.4. The simulated fits to the experimental $\Phi_T(B)/\Phi_T(0)$ curves are plotted in Fig. 4.4 in blue.

While the positions and heights of the resonance peaks in the triplet yields of both wires are captured well by these calculations, there is very poor agreement between experiment and theory in the high field region. We believe that this discrepancy suggests that the reaction scheme in Fig. 4.1 is incomplete; specifically, it is missing a magnetic field independent "background" contribution to the yield of ^{T}D–B–A*. In the absence of such a background, one would expect the high field triplet yield to be approximately one third of its zero-field value. This is a result of the high field effect described in Sect. 1.3.1: two of the three triplet states have a much larger energy gap to the singlet state in the high field limit than when there is no applied field, so intersystem crossing to those states will be significantly reduced. This argument is consistent with the raw simulations of the relative triplet yields in Fig. 4.4, which are tending towards a value of around 1/3 in the high field limit. However, it is clearly inconsistent with experimentally measured relative triplet yields.

Table 4.5 The singlet and triplet recombination rate constants of the $PTZ^{\bullet+}$–Ph_n–$PDI^{\bullet-}$ molecular wires

n	k_T / μs^{-1}	k_S / μs^{-1}
2	27500	47.4
3	3800	2.16
4	350	2.45
5	60.0	2.89

The discrepancies between our simulations and the experiments are resolved if we assume that the experiments are detecting an additional contribution to the triplet yield produced by some field independent process outside of the mechanism outlined in Fig. 4.1. This would in effect add a background contribution to the triplet yields calculated using the radical pair model. To account for this contribution, we have recalculated the relative triplet yield (RTY) as

$$\mathrm{RTY}(B) = \frac{\Phi_\mathrm{T}(B) + x}{\Phi_\mathrm{T}(0) + x}, \tag{4.3}$$

where $\Phi_\mathrm{T}(B)$ is the simulated triplet yield at magnetic field strength B, and x is the background contribution, defined as

$$x = \frac{\lambda\, \Phi_\mathrm{T}(0) - \Phi_\mathrm{T}(\infty)}{1 - \lambda}, \tag{4.4}$$

where λ is the experimental high field limit of the triplet yield.

New best fits to the experimental RTYs were found for both the $n = 2$ and 3 wires by using Eq. (4.3) to reoptimise the triplet recombination rate k_T. These fits are plotted in red in Fig. 4.4. With the background correction included, the simulations agree quantitatively with the experiments, allowing us to extract the optimum values of k_T and hence also k_S for both molecular wires. These are listed in Table 4.5. The empirical parameters used for the background corrections were $x = 0.0416$ and $\lambda = 0.9465$ when $n = 2$, and $x = 0.5253$ and $\lambda = 0.8136$ when $n = 3$. An *a posteriori* justification for including the background correction will be presented in Sect. 4.4.2, and several possible mechanisms by which the $^\mathrm{T}$D–B–A* state could be produced will be discussed in Sect. 4.4.3.

4.3.2 Longer Wires

For the wires with $n = 4$ and 5, Weiss et al. measured the radical pair yield 50 ns after the initial photoexcitation as a function of the strength of the applied magnetic field, and again reported this as the relative radical pair yield, $\Phi_\mathrm{RP}(B)/\Phi_\mathrm{RP}(0)$. In our simulations, we found a best fit to the experimental data at $t = 55$ ns, most likely due to the finite (7 ns) experimental instrument response time [16]. Our results are plotted

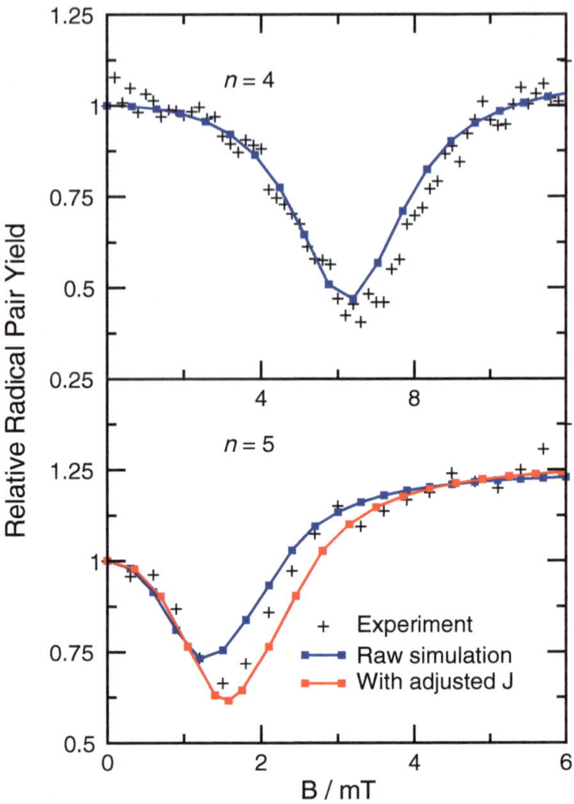

Fig. 4.5 The radical pair yield of $PTZ^{\bullet+}$–Ph_n–$PDI^{\bullet-}$ as a function of the strength of the applied magnetic field, relative to the radical pair yield in the absence of a field, for $n = 4$ above and $n = 5$ below. The blue curves were obtained using the exchange coupling constants given in Ref. [16]; the red curve in the lower panel is obtained with $2J/|\gamma_e| = 1.75$ mT, which has been fitted to the experimental peak position. The experimental data is taken from Ref. [16]

in Fig. 4.5 in blue. For $n = 4$, excellent quantitative agreement is observed between simulation and experiment without the need for any further correction. However, the results for $n = 5$ are not as good, with the simulated minimum in the radical pair yield at a different magnetic field strength than that observed experimentally.

When $n = 5$, the exchange coupling is comparable to the sum of the effective hyperfine fields of the two radicals (0.96 mT in $PTZ^{\bullet+}$ and 0.27 mT in $PDI^{\bullet-}$). As a result, intersystem crossing to the $|T_+\rangle$ state cannot entirely be neglected on resonance, as it can for the shorter wires. As the magnetic field strength is increased towards $2J/|\gamma_e|$, the energy gap between the $|S\rangle$ and $|T_+\rangle$ states increases, reducing the rate of transition between the two states. At the same time, the rate of crossing from $|S\rangle$ to $|T_-\rangle$ increases, becoming most efficient when $B = 2J/|\gamma_e|$. Therefore, the total intersystem crossing is most efficient, and a minimum in the radical pair yield is observed, at a field strength somewhat below $2J/|\gamma_e|$.

As a result of this, one cannot simply read off the magnitude of the exchange coupling constant J from the magnetic field strength at the minimum in the radical pair yield for the $n = 5$ wire. Instead, J must be extracted from a spin dynamics calculation. In order to do this, we varied J until the position of the minimum in the computed radical pair yield matched the experimental data.[2] Using the optimised value of J ($2J/|\gamma_e| = 1.75$ mT), a new lifetime-constrained (k_S, k_T) parameter space was constructed, defined by $k_S = ak_T^2 + bk_T + c$ with the constants $a = 6.523 \times 10^{-4}\,\mu s$, $b = -0.147$, and $c = 9.377\,\mu s^{-1}$, and from this a best fit to the experimental data was obtained by varying k_T. This best fit is shown as the red curve in the lower panel Fig. 4.5, and is in much better agreement with experiment than the raw simulation. The optimised rate constants for the longer wires are given along with those of the short wires in Table 4.5.

4.4 Discussion

4.4.1 Resonance Peak Widths

These simulations allow us to investigate a number of questions about the physics of the charge recombination along the $PTZ^{\bullet+}$–Ph_n–$PDI^{\bullet-}$ molecular wires. Firstly, why are the resonance peaks in the triplet and radical pair yield MFEs in Figs. 4.4 and 4.5 so broad? Previous experimental studies of similar D–B–A wires have noted that the width of the MFE peaks are often far greater than the size of the hyperfine interactions in the radical pair [31, 32]. That is also clear here: the sum of the effective hyperfine fields of the two radical is 1.23 mT, far smaller than the widths of peaks in the triplet yield MFE of ~100 mT and ~25 mT for the $n = 2$ and $n = 3$ wires respectively. Therefore, it seems implausible that hyperfine interactions alone could be responsible for the peak widths observed.

Our calculations suggest that the resonance peak widths are dominated by the lifetime broadening of the triplet state of the radical pair. Dividing the triplet recombination rate for the wires with $n = 2$ and 3 by the gyromagnetic ratio of an electron gives magnetic field strengths of 156 and 22 mT respectively, which are of the same order of magnitude as the observed resonance widths. The short lifetime of the triplet states of the radical pair leads to a broadening of their energy levels, giving a non-zero density of triplet states at the energy of the singlet state over a wide range of magnetic field strengths around the resonance condition $B = 2J/|\gamma_e|$. The singlet states of these radical pairs have much longer lifetimes, and so we do not expect lifetime broadening to have a significant effect on their density of states.

Figure 4.6 supports this explanation: for the $n = 3$ wire, the full width at half maximum (FWHM) of the peak in the simulated triplet yield increases monotonically with increasing k_T. Since the triplet recombination rate decreases as n increases, it is likely that when $n = 4$ the hyperfine interactions will contribute to the width of the radical pair yield resonance along with lifetime broadening, and when $n = 5$ they may well be the primary cause of the resonance width.

[2] We found that the position of the minimum in the radical pair yield was independent of the rate constants k_S and k_T.

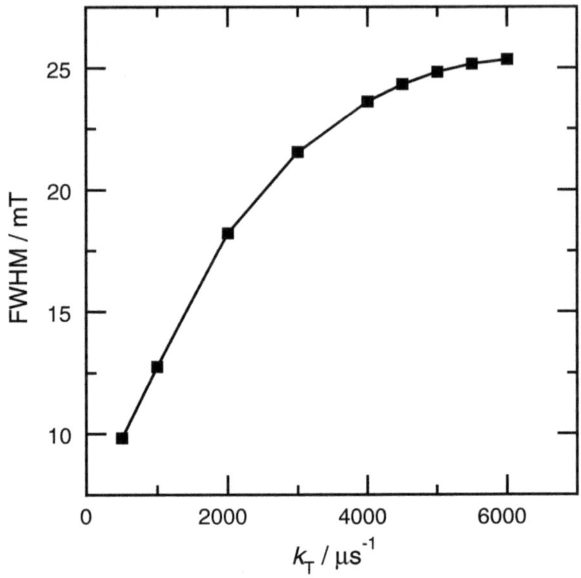

Fig. 4.6 The full width at half maximum of the simulated resonance peak in the triplet yield of $PTZ^{\bullet+}$–Ph_3–$PDI^{\bullet-}$ as a function of the triplet recombination rate, k_T. For each point, k_S is chosen to give the correct zero-field lifetime of the radical pair

4.4.2 Recombination Mechanisms

Secondly, what is the mechanism of charge recombination along the singlet and triplet pathways? Fig. 4.7 shows how k_S and k_T vary as a function of the distance between the radicals in the pair, with the rates plotted on a logarithmic scale. The triplet recombination rate decreases exponentially with the radical separation, with a decay constant $\beta_{TR} = 0.48$ Å^{-1}. This is very similar to that observed by Weiss et al. for the initial charge separation, $\beta_{CS} = 0.46$ Å^{-1} [16]. This exponential dependence is characteristic of the superexchange mechanism of recombination, indicating that the recombination of the triplet states of all four molecular wires occurs by superexchange.

The triplet recombination rate constants in Fig. 4.7 provide an *a posteriori* justification for introducing the background contribution to the triplet yield in Sect. 4.3.1. The values of k_T extracted from the simulations of the $n = 2$ and 3 wires which required the inclusion of a background are entirely consistent with those found for the $n = 4$ and 5 wires, which were obtained independently and without any need for a background. The quality of the single exponential fit to all four k_T data points in Fig. 4.7 ($R^2 = 0.997$) certainly supports this.

The singlet recombination rate constants in Fig. 4.7 are very similar for the $n = 3 - 5$ wires, but k_S is significantly larger when $n = 2$. This suggests that a change in the mechanism of the singlet recombination pathway occurs as the bridge length increases, with the superexchange mechanism most important when $n = 2$, while for the longer wires the incoherent hopping mechanism dominates.

Fig. 4.7 The singlet and triplet recombination rate constants k_S and k_T of the $PTZ^{\bullet+}$–Ph_n–$PDI^{\bullet-}$ radical pair for $n = 2 - 5$ extracted from our simulations, plotted as a function of the radical pair separation in those wires. k_T follows a single exponential with decay constant $\beta = 0.48$ Å^{-1}. The overall recombination rates k_{CR} are the reciprocals of the radical pair lifetimes in the absence of a magnetic field given in Table 4.3. These are taken from Ref. [16], along with the donor-acceptor distances

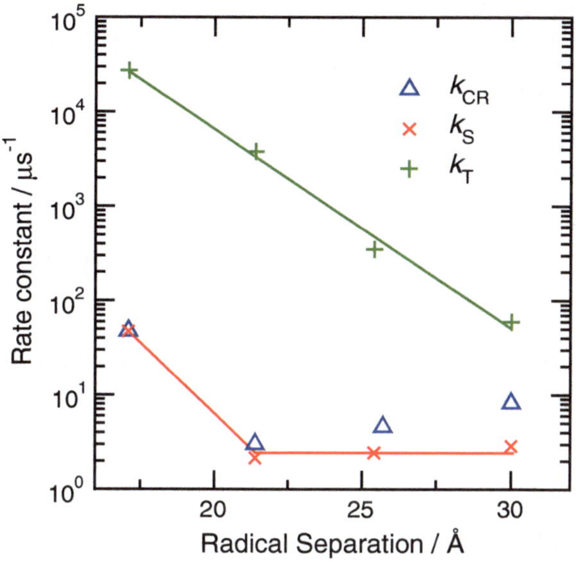

This change of mechanism can be understood in terms of Marcus theory. The direct recombination of the singlet radical pair lies deep in the Marcus inverted region [33], disfavouring the superexchange mechanism [16]. For wires with short bridges, the large electronic coupling between the electron donor and acceptor can compensate for this, so that when $n = 2$ charge recombination occurs primarily by the superexchange mechanism. However, as the bridge length increases and the electronic coupling decreases, the superexchange mechanism becomes slow compared to the incoherent hopping mechanism. On the other hand, direct recombination of the triplet radical pair is not as deep in the inverted region, because the triplet product is higher in energy than the singlet product [16]. As a result, the superexchange mechanism is far more favourable for the triplet radical pairs, and dominates for all bridge lengths.

These insights could not have been obtained from the overall experimental recombination rate constants k_{CR} alone, since they depend not only on k_S and k_T but also the rate of intersystem crossing between the spin states of the radical pair. This can be clearly seen by comparing k_S and k_{CR} for the $n = 2$ and $n = 5$ wires in Fig. 4.7. When $n = 2$, the exchange coupling is large, so intersystem crossing from the singlet state to the triplet state is very slow, and $k_{CR} \approx k_S$. However, when $n = 5$ the exchange coupling is much smaller and intersystem crossing is much more efficient, so recombination of the triplet radical pair contributes significantly to the experimental recombination rate and $k_{CR} > k_S$. Furthermore, k_{CR} will depend on the external magnetic field, whereas k_S and k_T do not. If the experimental measurements of k_{CR} were repeated in the presence of a magnetic field, we would expect them to satisfy $k_T > k_{CR}(B) > k_S$ for all magnetic field strengths B, with k_{CR} approaching k_T most closely at the resonant field strength, $B = 2J/|\gamma_e|$. However, $k_{CR}(B)$ will not actually reach k_T, so it is hard to see how k_S and k_T could be determined experimentally

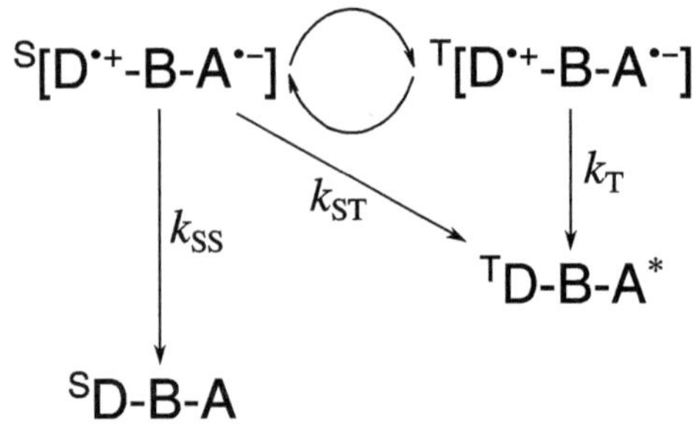

Fig. 4.8 A recombination reaction scheme including the possibility of intersystem crossing accompanying charge recombination. Here, k_{SS} and k_{ST} are the rate constants for recombination of the singlet radical pair to the singlet and triplet product states respectively

in this way. Instead, these rate constants must be determined by fitting the magnetic field dependence of the experimental results to quantum spin dynamics simulations, as we have done here.

4.4.3 The Origin of the Background

The final question raised by these results is that of the origin of the field independent background contribution to the triplet yield of the recombination reaction, which we introduced in Sect. 4.3.1. Here we will discuss four possible explanations of this background and their limitations.

Firstly, the triplet product could be generated by direct intersystem crossing from the excited singlet state,

$$^{S}\text{PTZ–Ph}_n\text{–PDI}^* \rightarrow {}^{T}\text{PTZ–Ph}_n\text{–PDI}^*. \tag{4.5}$$

It also possible that the triplet state could be formed by singlet fission,

$$^{S}\text{PTZ–Ph}_n\text{–PDI} + {}^{S}\text{PTZ–Ph}_n\text{–PDI}^* \rightarrow 2\,{}^{T}\text{PTZ–Ph}_n\text{–PDI}^*, \tag{4.6}$$

since the energy gap from the ground state to the excited triplet state is approximately half of the energy gap to the excited singlet state [16]. However, both of these processes appear to be ruled out by the control experiments of Weiss et al. which found the fluorescence yield of the compounds Ph_n–PDI to be 1, indicating that only charge separation competes with fluorescence in $^{S}\text{PTZ–Ph}_n\text{–PDI}^*$. It is perhaps possible that direct intersystem crossing might be promoted by the presence of a heavy sulphur atom in PTZ, but this seems unlikely if the electronic excitation in $^{S}\text{PTZ–Ph}_n\text{–PDI}^*$ is confined to the PDI chromophore.

A third possibility is intersystem occurring simultaneously with charge recombination,

$$^{S}[\text{PTZ}^{•+}\text{–Ph}_n\text{–PDI}^{•-}] \rightarrow {}^{T}\text{PTZ–Ph}_n\text{–PDI}^*. \tag{4.7}$$

which has been observed in these molecular wires [34]. This process is shown in radical pair reaction scheme in Fig. 4.8, labelled with the rate constant k_{ST} to distinguish it from 'normal' recombination of the singlet radical pair to the ground state, now labelled k_{SS}. Defining $f = k_{ST}/(k_{ST} + k_{SS})$ as the fraction of the singlet radical pair which reacts to form the triplet product, the total triplet yield from this scheme is

$$\begin{aligned} \Phi'_T(B) &= \Phi_T(B) + f\Phi_S(B) \\ &= (1-f)\Phi_T(B) + f, \end{aligned} \tag{4.8}$$

where Φ_T is still defined by Eq. (4.1). Comparing Eqs. (4.8)–(4.3), it is clear that

$$f = \frac{x}{x+1}, \tag{4.9}$$

and therefore that this mechanism will have the same effect as adding a field independent background contribution to the triplet yield. However, it is shown in Ref. [34] that the process in Eq. (4.7) only occurs at temperatures below 200 K in the molecular wires PTZ–Ph_n–PDI with $n \geq 2$, whereas the experiments of Weiss et al. that we have reproduced in this chapter were performed at room temperature [32].

Finally, the background could be due to a fraction of the radical pairs being formed in the triplet state, rather than the singlet state, during the initial charge separation:

$$^{S}\text{PTZ–Ph}_n\text{–PDI}^* \rightarrow {}^{T}[\text{PTZ}^{\bullet+}\text{–Ph}_n\text{–PDI}^{\bullet-}]. \tag{4.10}$$

We have checked in our simulations that the triplet recombination rates in these wires are sufficiently fast that these triplet radical pairs would recombine before any significant conversion to the singlet radical pair occurs, adding a field independent background contribution to the triplet yield of the radical pair recombination reaction. However, the fraction of radical pairs formed in the triplet state by this mechanism would need to be $f = x/(x+1) = 0.04$ when $n = 2$ and 0.34 when $n = 3$ in order to explain our results. The first of these fractions is consistent with the amount of intersystem crossing observed during charge separation in other radical pair reactions [35], but the second is much larger than would be expected on the basis of previous experiments. While the intersystem crossing will be promoted by the spin-orbit coupling associated with the sulphur atom in the PTZ radical, we can see no reason why the fraction of radical pairs formed in the triplet state would increase upon changing the bridge length from 2 to 3 para-phenylene units.

In summary, while we believe that some field independent background is required to explain the experimental triplet yields in Fig. 4.4, as discussed in Sect. 4.3.1 and justified *a posteriori* in Sect. 4.4.2, for now we remain unconvinced by all of the mechanisms we have proposed to account for this. It would be interesting if further experiments could be done to shed light on these mechanisms in an attempt to resolve this issue.

4.5 Conclusions

In this chapter, we have used the stochastic quantum mechanical method described in Sect. 2.5 to reproduce the magnetic field effects on the triplet and radical pair yields measured by Weiss et al. [16] for a series of $PTZ^{\bullet+}$–Ph_n–$PDI^{\bullet-}$ molecular wires with increasing bridge lengths $n = 2 - 5$. We have extracted recombination rates for the singlet and triplet states of the radical pair from our simulations, and used them to shed light on the spin dynamics and charge recombination mechanisms of these molecular wires. The very wide peaks in the triplet yield MFEs observed experimentally and reproduced in our simulations are the result of lifetime broadening of the triplet states. The triplet rates follow a single exponential decay as a function of radical separation, consistent with the superexchange recombination mechanism. By contrast, the singlet rates are very similar for wires with bridges consisting of three or more p-phenylene rings, suggesting that incoherent hopping is the primary recombination mechanism in the singlet pathway. The difference between the mechanisms of the two pathways can been explained using Marcus theory. The fact that charge transport along the singlet and triplet states of the molecular wires operates via different mechanisms precludes a straightforward rationalisation of the change in radical pair lifetimes with bridge length. Instead, the separate singlet and triplet rates must be extracted in order to understand the behaviour of these wires. Finally, we have found good evidence for a magnetic field independent background contribution to the triplet yield of the charge recombination reaction, the physical origin of which remains an open question.

References

1. Nitzan, A. (2001). Electron transmission through molecules and molecular interfaces. *Annual Review of Physical Chemistry, 52*, 681–750.
2. Weiss, E. A., et al. (2005). Conformationally gated switching between superexchange and hopping within oligo-p-phenylene-based molecular wires. *Journal of the American Chemical Society, 127*, 11842–11850.
3. Goldsmith, R. H., et al. (2005). Wire-like charge transport at near constant bridge energy through uorene oligomers. *Proceedings of the National Academy of Sciences of the United States of America, 102*, 3540–5.
4. Fromme, P. (2002). Functional implications on the mechanism of the function of photosystem II including water oxidation based on the structure of photosystem II. Philosophical Transactions of the Royal Society of London. *Series B, Biological Sciences, 357*, 1337–44.
5. Parson, W. W. (2003). Electron donors and acceptors in the initial steps of photosynthesis in purple bacteria: a personal account. *Photosynthesis Research, 76*, 81–92.
6. Wasielewski, M. R. (2006). Energy, charge, and spin transport in molecules and self-assembled nanostructures inspired by photosynthesis. *Journal of Organic Chemistry, 71*, 5051–5066.
7. Jortner, J., & Ratner, M. A. (1997). *Molecular Electronics*. Oxford: Blackwell Science.
8. Davis, W. B., Wasielewski, M. R., Ratner, M. A., Mujica, V., & Nitzan, A. (1997). Electron transfer rates in bridged molecular systems: a phenomenological approach to relaxation. *The Journal of Physical Chemistry A, 101*, 6158–6164.
9. Davis, W. B., Ratner, M. A., & Wasielewski, M. R. (2002). Dependence of electron transfer dynamics in wire-like bridge molecules on donor-bridge energetics and electronic interactions. *Chemical Physics, 281*, 333–346.

10. Moser, C. C., Keske, J. M., Warncke, K., Farid, R. S., & Dutton, P. L. (1992). Nature of biological electron transfer. *Nature, 355*, 796–802.
11. Farid, R. S., Moser, C. C., & Dutton, P. L. (1993). Electron transfer in proteins. *Current Opinion in Structural Biology, 3*, 225–233.
12. Mikkelsen, K. V., & Ratner, M. A. (1987). Electron tunneling in solid-state electron-transfer reactions. *Chemical Reviews, 87*, 113–153.
13. Barbara, P. F., Meyer, T. J., & Ratner, M. A. (1996). Contemporary issues in electron transfer research. *The Journal of Physical Chemistry, 100*, 13148–13168.
14. Murphy, C. J., et al. (1994). Fast photoinduced electron transfer through DNA intercalation. *Proceedings of the National Academy of Sciences of the United States of America, 91*, 5315–9.
15. Davis, W. B., Svec, W. A., Ratner, M. A., & Wasielewski, M. R. (1998). Molecular-wire behaviour in p-phenylenevinylene oligomers. *Nature, 396*, 60–63.
16. Weiss, E. A. (2004). Making a molecular wire: charge and spin transport through. *Journal Of The American Chemical Society, 126*, 5577–5584.
17. Jang, S., Berkelbach, T. C., & Reichman, D. R. (2013). Coherent quantum dynamics in donor-bridge-acceptor systems: beyond the hopping and super-exchange mechanisms. *New Journal of Physics, 15*, 105020.
18. Anderson, P. W. (1959). New approach to the theory of superexchange interactions. *Physical Review, 115*, 2–13.
19. McConnell, H. M. (1961). Intramolecular charge transfer in aromatic free radicals. *Journal of Chemical Physics, 35*, 508–515.
20. Jortner, J., Bixon, M., Langenbacher, T., & Michel-Beyerle, M. E. (1998). Charge transfer and transport in DNA. *Proceedings of the National Academy of Sciences of the United States of America, 95*, 12759–12765.
21. Berlin, Y. A., Burin, A. L., & Ratner, M. A. (2002). Elementary steps for charge transport in DNA: thermal activation versus tunneling. *Chemical Physics, 275*, 61–74.
22. Steiner, U. E., & Ulrich, T. (1989). Magnetic field effects in chemical kinetics and related phenomena. *Chemical Reviews, 89*, 51–147.
23. Chiu, M. F., Gilbert, B. C. & Hanson, P. A study by electron spin resonance of some heterocyclic radicals containing elements of group VI. *Journal of the Chemical Society B: Physical Organic*, 1700 (1970).
24. Tauber, M. J., Kelley, R. F., Giaimo, J. M., Rybtchinski, B., & Wasielewski, M. R. (2006). Electron hopping in pi-stacked covalent and self-assembled perylene diimides observed by ENDOR spectroscopy. *Journal of the American Chemical Society, 128*, 1782–1783.
25. Lee, C., Yang, W., & Parr, R. G. (1988). Development of the Colle-Salvetti correlation-energy formula into a functional of the electron density. *Physical Review B, 37*, 785–789.
26. Becke, A. D. (1993). A new mixing of Hartree-Fock and local density-functional theories. *The Journal of Chemical Physics, 98*, 1372.
27. Barone, V. (1995). *Recent Advances in Density Functional Methods: Part I*. Singapore: World Scientific.
28. Dunning, T. H, Jr. (1989). Gaussian basis sets for use in correlated molecular calculations. I. the atoms boron through neon and hydrogen. *The Journal of Chemical Physics, 90*, 1007.
29. Woon, D. E., & Dunning, T. H, Jr. (1993). Gaussian basis sets for use in correlated molecular calculations. III. the atoms aluminum through argon. *Journal of Chemical Physics, 98*, 1358.
30. Hasharoni, K. (1995). Mimicry of the radical pair and triplet-states in photosynthetic reaction centers with a synthetic model. *Journal of the American Chemical Society, 117*, 8055–8056.
31. Weiss, E. A., Ratner, M. A., & Wasielewski, M. R. (2003). Direct measurement of singlet - triplet splitting within rodlike photogenerated radical ion pairs using magnetic field effects: estimation of the electronic coupling for charge recombination. *Journal of Physical Chemistry, 107*, 3639–3647.
32. Weiss, E. A., Tauber, M. J., Ratner, M. A., & Wasielewski, M. R. (2005). Electron spin dynamics as a probe of molecular dynamics: Temperature-dependent magnetic field effects on charge recombination within a covalent radical ion pair. *Journal of the American Chemical Society, 127*, 6052–6061.

33. Marcus, R. A. (1965). On the theory of electron-transfer reactions. VI. unified treatment for homogeneous and electrode reactions. *Journal of Chemical Physics*, *43*, 679.
34. Dance, Z. E. X., et al. (2006). Time-resolved EPR studies of photogenerated radical ion pairs separated by p-phenylene oligomers and of triplet states resulting from charge recombination. *Journal of Physical Chemistry B*, *110*, 25163–25173.
35. Maeda, K., et al. (2011). Spin-selective recombination kinetics of a model chemical magnetoreceptor. *Chemical Communications*, *47*, 6563–6565.

Chapter 5
Avian Magnetoreception

In 2000, Ritz and Schulten proposed that a radical pair reaction could be responsible for the magnetoreception observed in some birds [1]. This suggestion gained further traction in 2008, when Maeda et al. showed that the recombination rate of a carotenoid-porphyrin-fullerene radical pair was affected by the application of an Earth-strength magnetic field [2]. At the same time, substantial circumstantial evidence for the involvement of a radical pair reaction in the avian compass was mounting. However, in order for such a reaction to act as a biological compass, it must have an *anisotropic* response to an Earth-strength magnetic field. This has not yet been observed experimentally, so theoretical studies of the cryptochrome-based radical pair thought to be responsible for magnetoreception are required to assess the likelihood of this mechanism being the basis of the magnetic compass of migratory birds.

In this chapter, we will review the evidence for the radical pair mechanism of magnetoreception, and discuss the possible identity of the radical pair responsible. We shall then use the semiclassical theory outlined in Sect. 3.1 to reproduce the results of Maeda et al.'s experiments which demonstrated an Earth-strength magnetic field effect on a radical pair reaction. These simulations also provide some insight into the effect of electron spin relaxation on this reaction. Having established an isotropic magnetic field effect, we conclude the chapter by investigating the plausibility of the cryptochrome radical pair exhibiting an anisotropic magnetic field effect in an Earth-strength field.[1]

[1] Parts of this chapter have been reproduced with permission from Lewis, A. M., Manolopoulos, D. E. & Hore, P. J. *Journal of Chemical Physics* **141**, 044111 (2014), https://aip.scitation.org/doi/abs/10.1063/1.4890659

A. Lewis, *Spin Dynamics in Radical Pairs*, Springer Theses,
https://doi.org/10.1007/978-3-030-00686-0_5

5.1 Background

The ability of birds to use the Earth's magnetic field as a navigational aid was first proposed in the 19th century [3, 4], and was observed conclusively in European robins in 1968 [5]. During their migration season, it was found that caged birds preferred to orient themselves in the direction of migration. When the direction of magnetic north was changed artificially, the birds reoriented themselves according to the new alignment of the magnetic field. Since then, it has been shown that many other species of bird exhibit magnetoreception [6], and that the compass in some migratory birds can provide very precise directional information, detecting the axis of magnetic field lines to within 5° [7, 8]. There are two main hypotheses for the mechanism of this magnetoreception: a radical pair model and a magnetite based model. We will discuss the evidence for the former in some detail, before briefly commenting on the latter.

5.1.1 The Radical Pair Mechanism

There are several features of the avian magnetic compass which are consistent with the radical pair mechanism of magnetoreception. Firstly, it is an inclination compass: it is independent of the polarity of the applied magnetic field, based instead on the axial direction of the field lines [9]. Therefore at the equator, when the field lines are parallel to the surface of the Earth, the compass provides no information; migratory birds become disoriented in a horizontal field [9]. Away from the equator, the compass does not distinguish between north and south, instead distinguishing between "polewards", following magnetic field lines angled towards the Earth, and "equatorwards", following magnetic field lines angled away from the Earth. This appears to present a problem for migratory birds which cross the equator, such as the garden warbler. However, it has been shown that exposure to a horizontal field acts as a trigger to change the warbler's direction of migration from equatorwards to polewards, so their migration continues in the correct direction after crossing the equator [10].

The singlet yield of a radical pair reaction, Φ_S, is also independent of the polarity of an applied magnetic field, as we shall now prove. Firstly, we must define the time reversal operator, $\hat{\Theta}$,

$$\hat{\Theta}\,|\Psi, t\rangle = |\Psi, -t\rangle\,. \tag{5.1}$$

Unusually for a quantum mechanical operator, $\hat{\Theta}$ is anti-linear [11]:

$$\hat{\Theta}\,(c_1\,|1\rangle + c_2\,|2\rangle) = c_1^*\,\hat{\Theta}\,|1\rangle + c_2^*\,\hat{\Theta}\,|2\rangle\,. \tag{5.2}$$

This has several interesting consequences; of particular significance here are the properties [11]

$$\hat{\Theta}\, i \hat{H} = -i\, \hat{\Theta} \hat{H},$$
$$\hat{\Theta}\, \hat{\mathbf{J}}\, \hat{\Theta}^{-1} = -\hat{\mathbf{J}}, \tag{5.3}$$

where $\hat{\mathbf{J}}$ is any angular momentum vector operator. We can apply the time reversal operator to the Hamiltonian in Eq. (2.1):

$$\begin{aligned}
\hat{\Theta}\, i \hat{H}(\mathbf{B})\, \hat{\Theta}^{-1} &= -i\, \hat{\Theta} \hat{H}(\mathbf{B}) \hat{\Theta}^{-1} \\
&= -i\, \hat{\Theta} \left[\sum_{i=1}^{2} \left(-\gamma_e \mathbf{B} \cdot \hat{\mathbf{S}}_i + \sum_{k=1}^{N_i} \hat{\mathbf{I}}_{ik} \cdot \mathbf{A}'_{ik} \cdot \hat{\mathbf{S}}_i \right) + \hat{\mathbf{S}}_1 \cdot \mathbf{D}' \cdot \hat{\mathbf{S}}_2 \right] \hat{\Theta}^{-1} \\
&= -i\, \hat{\Theta} \left[\sum_{i=1}^{2} \left(-\gamma_e \mathbf{B} \cdot \hat{\mathbf{S}}_i + \sum_{k=1}^{N_i} \hat{\mathbf{I}}_{ik} \hat{\Theta}^{-1} \cdot \mathbf{A}'_{ik} \cdot \hat{\Theta} \hat{\mathbf{S}}_i \right) \right. \\
&\qquad \left. + \hat{\mathbf{S}}_1 \hat{\Theta}^{-1} \cdot \mathbf{D}' \cdot \hat{\Theta} \hat{\mathbf{S}}_2 \right] \hat{\Theta}^{-1} \\
&= -i \left[\sum_{i=1}^{2} \left(+\gamma_e \mathbf{B} \cdot \hat{\mathbf{S}}_i + \sum_{k=1}^{N_i} \hat{\mathbf{I}}_{ik} \cdot \mathbf{A}'_{ik} \cdot \hat{\mathbf{S}}_i \right) + \hat{\mathbf{S}}_1 \cdot \mathbf{D}' \cdot \hat{\mathbf{S}}_2 \right] \\
&= -i\, \hat{H}(-\mathbf{B}),
\end{aligned} \tag{5.4}$$

where we have used the results in Eq. (5.3), and twice inserted the unit operator $\hat{1} = \hat{\Theta}^{-1} \hat{\Theta}$. We may also apply the time reversal operator to the singlet projection operator

$$\begin{aligned}
\hat{\Theta} \hat{P}_{\mathrm{S}}\, \hat{\Theta}^{-1} &= \hat{\Theta} \left(\frac{1}{4} \hat{1} - \hat{\mathbf{S}}_1 \hat{\Theta}^{-1} \cdot \hat{\Theta} \hat{\mathbf{S}}_2 \right) \hat{\Theta}^{-1} \\
&= \left(\frac{1}{4} \hat{1} - \hat{\mathbf{S}}_1 \cdot \hat{\mathbf{S}}_2 \right) = \hat{P}_{\mathrm{S}},
\end{aligned} \tag{5.5}$$

and by doing so establish that the recombination operator is unaffected by application of the time reversal operator:

$$\begin{aligned}
\hat{\Theta} \hat{K}\, \hat{\Theta}^{-1} &= \hat{\Theta} \left(\frac{k_{\mathrm{S}}}{2} \hat{P}_{\mathrm{S}} + \frac{k_{\mathrm{T}}}{2} \hat{P}_{\mathrm{T}} \right) \hat{\Theta}^{-1} \\
&= \hat{\Theta} \left(\frac{k_{\mathrm{S}}}{2} \hat{P}_{\mathrm{S}} + \frac{k_{\mathrm{T}}}{2} \left(\hat{1} - \hat{P}_{\mathrm{S}} \right) \right) \hat{\Theta}^{-1} \\
&= \left(\frac{k_{\mathrm{S}}}{2} \hat{P}_{\mathrm{S}} + \frac{k_{\mathrm{T}}}{2} \left(\hat{1} - \hat{P}_{\mathrm{S}} \right) \right) = \hat{K}.
\end{aligned} \tag{5.6}$$

The quantum mechanical definition of the singlet probability of a radical pair initially in the singlet state is given by Eqs. (2.42) and (2.43), with $\hat{A} = \hat{\rho}(0) = \hat{P}_{\mathrm{S}}$:

$$\mathrm{P}_{\mathrm{S}}(\mathbf{B}, t) = \frac{1}{Z} \mathrm{tr} \left[\hat{P}_{\mathrm{S}}\, e^{+i \hat{H}(\mathbf{B}) t - \hat{K} t}\, \hat{P}_{\mathrm{S}}\, e^{-i \hat{H}(\mathbf{B}) t - \hat{K} t} \right]. \tag{5.7}$$

We can use Eqs. (5.4) and (5.6) to find the effect of applying the time reversal operator to the propagation operator,

$$\begin{aligned}
\hat{\Theta}\, e^{\pm i\hat{H}(\mathbf{B})t-\hat{K}t}\, \hat{\Theta}^{-1} &= \lim_{n\to\infty} \hat{\Theta}\left[1 \pm \frac{it}{n}\hat{H}(\mathbf{B}) - \frac{t}{n}\hat{K}\right]^n \hat{\Theta}^{-1} \\
&= \lim_{n\to\infty} \left[\hat{\Theta}\left(1 \pm \frac{it}{n}\hat{H}(\mathbf{B}) - \frac{t}{n}\hat{K}\right)\hat{\Theta}^{-1}\right]^n \\
&= \lim_{n\to\infty} \left[1 \mp \frac{it}{n}\hat{H}(-\mathbf{B}) - \frac{t}{n}\hat{K}\right]^n \\
&= e^{\mp i\hat{H}(-\mathbf{B})t-\hat{K}t}.
\end{aligned} \tag{5.8}$$

Then, inserting $\hat{1} = \hat{\Theta}^{-1}\hat{\Theta}$ into Eq. (5.7) several times and using the results in Eqs. (5.5) and (5.8) along with the invariance of a trace to cyclic permutation, we find

$$\begin{aligned}
\mathrm{P_S}(\mathbf{B}, t) &= \frac{1}{Z}\mathrm{tr}\left[\hat{P}_S\, \hat{\Theta}^{-1}\hat{\Theta}\, e^{+i\hat{H}(\mathbf{B})t-\hat{K}t}\, \hat{\Theta}^{-1}\hat{\Theta}\hat{P}_S\, \hat{\Theta}^{-1}\hat{\Theta}e^{-i\hat{H}(\mathbf{B})t-\hat{K}t}\, \hat{\Theta}^{-1}\hat{\Theta}\right] \\
&= \frac{1}{Z}\mathrm{tr}\left[\hat{P}_S\, \hat{\Theta}^{-1}\, e^{-i\hat{H}(-\mathbf{B})t-\hat{K}t}\, \hat{\Theta}\hat{P}_S\, \hat{\Theta}^{-1}e^{+i\hat{H}(-\mathbf{B})t-\hat{K}t}\, \hat{\Theta}\right] \\
&= \frac{1}{Z}\mathrm{tr}\left[\hat{\Theta}\hat{P}_S\, \hat{\Theta}^{-1}e^{+i\hat{H}(-\mathbf{B})t-\hat{K}t}\, \hat{\Theta}\hat{P}_S\hat{\Theta}^{-1}\, e^{-i\hat{H}(-\mathbf{B})t-\hat{K}t}\right] \\
&= \frac{1}{Z}\mathrm{tr}\left[\hat{P}_S\, e^{+i\hat{H}(-\mathbf{B})t-\hat{K}t}\, \hat{P}_S\, e^{-i\hat{H}(-\mathbf{B})t-\hat{K}t}\right] \\
&= \mathrm{P_S}(-\mathbf{B}, t).
\end{aligned} \tag{5.9}$$

It then follows immediately from Eq. (5.9) and the definition of the singlet yield in Eq. (2.60) that

$$\Phi_S(\mathbf{B}) = \Phi_S(-\mathbf{B}), \tag{5.10}$$

as we set out to prove. Hence radical pair reactions are also sensitive only to the orientation of an applied magnetic field, and not its polarity.

Secondly, the avian compass is very sensitive to the strength of the magnetic field [12]. Birds could not orient themselves in a magnetic field strength even 10 μT different from that which they were acclimatised to. However, they were able to orient themselves once they had been exposed to the new field strength for three days. This can be understood within the radical pair hypothesis – the form of the magnetic field effect will depend on the applied field strength, so birds may take time to adapt to a new compass pattern [13].

Thirdly, the birds' ability to orient themselves has been shown to depend on the wavelength of the ambient light. The magnetic compass has been shown to operate under blue and green light but not under red light [14]. This observation is easily interpreted within the radical pair hypothesis: the energy threshold which must be reached for successful photoexcitation and subsequent radical pair formation occurs at shorter wavelengths than red light [1]. However, it has been pointed out that the light-dependence of orientation may in fact arise from a separate process which affects the birds' motivation to act upon information provided by a chemical compass, rather than magnetoreception itself being light dependent [13, 15].

Perhaps the key observations which support the radical pair hypothesis are the effects of radio frequency (RF) fields on avian magnetoreception. The rate and yields of radical pair reactions are expected to change when exposed to RF fields, since the fields will induce transitions when on resonance with the Zeeman or hyperfine splittings between the spin states of the radical pair and hence modulate the singlet-triplet conversion rate [1, 16]. This phenomenon has been demonstrated in at least two organic radical pairs [17, 18], leading to the prediction that if avian magnetoreception did operate by the radical pair mechanism, birds would be disoriented in the presence of RF fields. This was indeed observed [19–21], and the experiments had three features of particular interest. Firstly, the birds were unable to adapt to the RF fields as they were to a change in the strength of the static magnetic field, suggesting the presence of these fields directly disables magnetoreception [22]. Secondly, even incredibly weak RF fields caused disorientation, suggesting an extremely long lifetime of the radical pair involved in magnetoreception [13, 21]. Finally, when the RF field was parallel to the magnetic field, the birds remained able to orient themselves. This implies that one of the radicals in the pair has no significant hyperfine interactions [2, 23]. However, this last claim has since been disputed [24].

There are a number of conditions which must be satisfied for an Earth-strength magnetic field effect to be observed in a radical pair reactions [1, 15]. To begin with, the recombination rate of the radical pair must be slow. Ritz suggested that an average rate of $\bar{k} < 1\ \mu\text{s}^{-1}$ is necessary for significant mixing of the singlet and triplet states to take place before recombination occurs in an Earth-strength magnetic field [1]. In addition, the radical pair must be essentially immobile. If the radicals rotate with a correlation time of $\leq 1\ \mu\text{s}$, any anisotropy in the reaction yield will be averaged out, and no directional information would be available to form a magnetic compass [13]. Furthermore, molecular motions modulate the hyperfine interactions and induce relaxation of the electron spins, destroying the correlation between them. Therefore, for radical pair reactions to act as magnetoreceptors the relaxation times must be longer than the radical pair lifetimes. Recent studies show that while some relaxation caused by limited molecular motion may be tolerated, at present it is difficult to reconcile the extremely long lifetimes implied by the RF field experiments with the relaxation inevitably caused by even small motions [25–27].

5.1.2 Cryptochrome

The radical pair most likely to be responsible for magnetoreception in European robins consists of a flavin adenide nucleotide (FAD) radical and a tryptophan (Trp_C or W_C) radical found in the cryptochrome protein [1, 13, 15], whose structure is shown on the left of Fig. 5.1. The relative orientation of the two radicals is taken

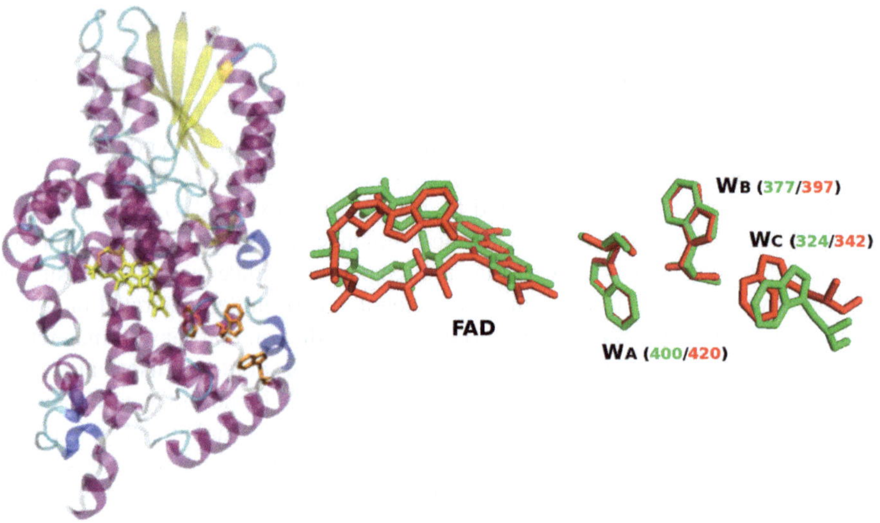

Fig. 5.1 The structure of the cryptochrome protein is displayed on the left, with the position of the FAD cofactor highlighted in yellow in the centre of the protein and the three Trp residues shown in orange to its right. On the right, the FAD cofactor and Trp residues are displayed in the crystallographic orientation for DmCry (red) and AtCry (green). It is the terminal tryptophan, W_C, which forms the long-lived radical pair along with FAD

from the crystal structure of the *Drosophila melanogaster* cryptochrome (DmCry) [28, 29]. This is shown in red on the right-hand side of Fig. 5.1, along with the relative positions of the two radicals in the *Arabidopsis thaliana* cryptochrome (AtCry), a plant cryptochrome, in green [30]. The crystal structure of the proposed avian magnetoreceptor, Cry1a, has not yet been elucidated, but the similarity between the cryptochrome structures which have been identified gives us confidence that the DmCry structure will serve as a suitable proxy for Cry1a.

The 15 magnetic nitrogen and hydrogen nuclei in the FAD radical and 12 magnetic nuclei in the Trp_C radical are shown in Fig. 5.2; their hyperfine tensors are listed in Appendix E. The radical pair is formed by the photoexcitation of FAD using blue light, which is followed by a series of rapid ($k_{et} > 10^8$ s^{-1}) [31, 32] electron transfers along a chain of three tryptophan residues resulting in the oxidation of the terminal Trp_C to create the long-lived radical pair [32, 33]. In solution, the recombination of this radical pair has been shown to be influenced by a magnetic field, suggesting its suitability as a magnetoreceptor [34, 35]. Furthermore, it survives long enough to be influenced by an Earth-strength magnetic field [34], and is found in high abundance in the cell membrane of photoreceptors in the retina of birds [36, 37]. Finally, the separation between the two radicals is sufficiently large that there will not be a significant energy gap between the singlet and triplet states of the radical pair, but the charge separation can still occur rapidly, since it takes place through three short-range electron transfer reactions [13].

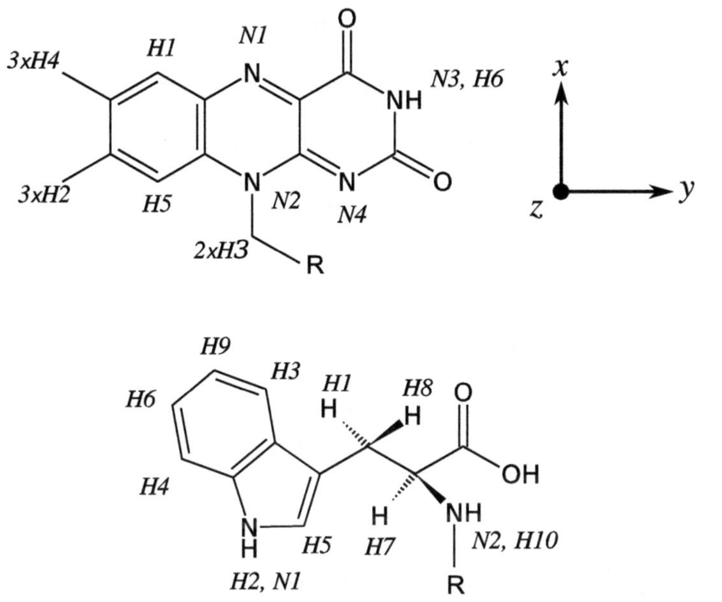

Fig. 5.2 The 15 magnetic nuclei in the flavin (FAD$^{\bullet-}$) radical are shown above, along with the definition of the x, y and z axes of the flavin radical used in the plots of the anisotropy of the singlet yield later in this chapter. The 12 magnetic nuclei in the tryptophan ($W_C^{\bullet+}$) radical are shown below. The anisotropic hyperfine interaction tensors of these nuclei are listed in Appendix E

However, one feature of this radical pair is not consistent with the experimental evidence described earlier. Both radicals have significant hyperfine interactions, which is inconsistent with the observation that RF fields parallel to the Earth's field do not disorient birds. Furthermore, while the FAD radical is almost ideally suited to magnetoreception [32, 38], pairing it with a radical with many hyperfine interactions like tryptophan is not [39, 40]. In fact, simulations have shown that replacing the tryptophan with a hypothetical radical with no hyperfine interactions increases the anisotropic response of the radical pair a hundredfold [38]. This has lead to speculation about alternative partners for the FAD radical in the pair, such as ascorbic acid [38] or a fully reduced FAD forming a radical pair with a superoxide molecule [41, 42], although the latter seems unlikely due to the extremely fast relaxation of the superoxide radical as a result of its spin-orbit coupling [43].

5.1.3 *The Magnetite Hypothesis*

An alternative hypothesis of magnetoreception contends that chains of magnetite particles, crystals of Fe_3O_4, are affected by the Earth's magnetic field and provide directional information for a biological compass. Particles as small as 50 nm possess a large enough permanent dipole moment that they will align with a 50 μT field [44, 45]. Chains of these particles are found in a number of bacteria which align themselves with and swim along an Earth-strength magnetic field [46]. By contrast to avian magnetoreception, this alignment is passive; killed bacteria continued to be aligned with an external field, but no longer moved along its field lines.

These particles are present in a wide variety of organisms, and so have been proposed as a possible mechanism for avian magnetoreception [47, 48]. However, magnetite chains have not been found in any species of bird, and the mere presence of magnetite does not necessarily imply its involvement in magnetoreception[37]. Furthermore, it is hard to see how magnetite would be affected by a RF field – the crystals are far too large to re-align themselves a million times a second – and therefore difficult to reconcile the magnetite mechanism of magnetoreception with the experimental observation that RF fields disorient birds [19, 49]. Instead, it is possible that magnetite is involved in a magnetic "map", rather than a magnetic compass [14, 32].

5.2 A Prototypical Magnetoreceptor

The first experiment which demonstrated that an Earth-strength magnetic field could influence the outcome of a radical pair reaction was performed by Maeda et al. on a carotenoid-porphyrin-fullerene (CPF) triad, which is shown in Fig. 5.3 along with its photochemical reaction scheme [2]. The radical pair $C^{\bullet+}PF^{\bullet-}$ is formed by photoexcitation of the porphyrin ring followed by two rapid electron transfers, firstly from the porphyrin to the fullerene, then from the carotenoid to the porphyrin. These electron transfers preserve the singlet character of the initial photoexcited state [50], since singlet-triplet conversion is inhibited by the large exchange interaction in the precursor excited states, so the radical pair is formed predominately in the singlet state. It then undergoes coherent intersystem crossing to the triplet state due to the hyperfine interactions between the electron and nuclear spins on the carotenoid radical. The singlet and triplet radical pairs recombine at different rates, k_S and k_T.

Maeda et al. followed the recombination of the radical pairs at 113 K using transient absorption spectroscopy, measuring the total (singlet and triplet) radical pair population as a function of time after the initial photoexcitation. The magnetic field effect on the reaction was then found by subtracting the transient absorption curve obtained in the absence of a field from those found in 39 and 49 μT magnetic fields. This measurement is proportional to the difference in the survival probability of the radical pair, $\mathbf{1}(B,t) - \mathbf{1}(0,t)$. Note that if the recombination rates were symmetric, this quantity would equal $e^{\bar{k}t} - e^{\bar{k}t} = 0$ at all times; the observation of a magnetic field effect on the survival probability is dependent on the asymmetry of the recombination rates.

Here, we will reproduce the experimental results presented in Ref. [2] with the semiclassical method described in Sect. 3.1, in which the survival probability $\mathbf{1}(B,t)$ is given by Eq. (3.14) where $\bar{A} = \bar{\mathbf{1}}$. This demonstrates the utility of our semiclassical

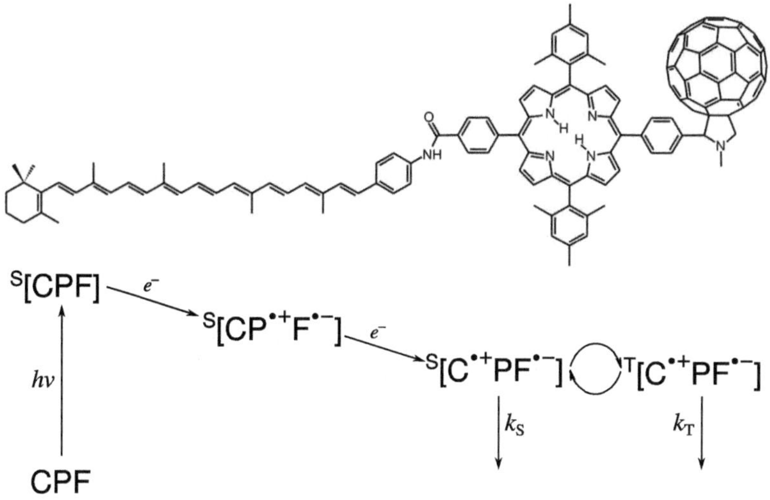

Fig. 5.3 The carotenoid-porphyrin-fullerene triad studied in Ref. [2], along with a diagram showing the photochemistry which precedes the coherent electron spin evolution and asymmetric recombination of the $C^{\bullet+}PF^{\bullet-}$ radical pair

method, since the problem is too complex to handle quantum mechanically: the carotenoid contains 45 magnetic hydrogen nuclei, and so the Hilbert space of the radical pair contains 1.4×10^{14} states. We will then use our approach to shed some light onto the underlying spin dynamics of the $C^{\bullet+}PF^{\bullet-}$ radical pair.

5.2.1 Simulation Details

In order to simulate this radical pair, several pieces of information are required: the singlet and triplet recombination rates, the fraction of radical pairs formed in the singlet state, which we will call λ, the hyperfine coupling constants, and the relaxation times of the electron spins on the two radicals. The first three have been determined from EPR experiments at 110 K to be $k_S = 1.8 \times 10^7$ s^{-1}, $k_T = 7.1 \times 10^4$ s^{-1}, and $\lambda = 0.93$ [50]. We shall assume that these values are unchanged at 113 K, the temperature at which the transient absorption experiments were performed. To account for the small initial population of triplet radical pairs, the classical density variable in Eq. (3.14) is given by[2]

$$\bar{\rho}(0) = \lambda \bar{P}_S(0) + \frac{1-\lambda}{3} \bar{P}_T(0). \tag{5.11}$$

[2]Note that any initial density of this form will still satisfy the condition in Eq. (5.10), since $\hat{P}_T = \hat{1} - \hat{P}_S$.

Table 5.1 The isotropic hyperfine coupling constants of the carotenoid cation radical, calculated using the B3LYP functional and the EPR-II basis set

k	$a_{1k}/\|\gamma_e\|$ (mT)	k	$a_{1k}/\|\gamma_e\|$ (mT)	k	$a_{1k}/\|\gamma_e\|$ (mT)	k	$a_{1k}/\|\gamma_e\|$ (mT)	k	$a_{1k}/\|\gamma_e\|$ (mT)
1	0.04879	10	−0.021817	19	0.480492	28	0.015707	37	0.018443
2	0.046328	11	−0.140593	20	0.015702	29	0.49151	38	0.001563
3	−0.115098	12	−0.087963	21	0.490845	30	0.407741	39	−0.017735
4	−0.111317	13	−0.071456	22	0.332346	31	0.17369	40	0.014287
5	−0.361254	14	0.050581	23	0.306659	32	0.579152	41	−0.028314
6	0.130081	15	−0.275215	24	0.011857	33	0.057321	42	0.003183
7	0.094903	16	0.056448	25	0.008375	34	0.006161	43	0.246942
8	−0.316911	17	0.111917	26	0.246503	35	−0.005099	44	0.442049
9	0.094676	18	−0.385563	27	0.257005	36	−0.003271	45	0.027488

There are 45 hydrogen nuclei with significant hyperfine interactions on the carotenoid, and none on the fullerene. We will assume that the electron and hole are localised on these two species, and so neglect the hydrogen and nitrogen nuclei on the porphyrin ring. The hyperfine interactions of the carotenoid radical cation ($CH^{\bullet +}$) can be obtained from a density functional theory calculation using the B3LYP functional [51, 52] and the EPR-II basis set [53], which is optimised for this type of problem, and are listed in Table 5.1. Terminating the carotenoid simply with a H atom ensures that the hole will indeed be localised on the carotenoid radical; it has been shown that the self-interaction problem leads to excessive delocalisation if the porphyrin linker is included in the DFT calculation [54]. Note that positions of the nuclei corresponding to each of these hyperfine interactions have no effect to the spin dynamics of the radical.

Finally, we must consider the relaxation of the spins in the radical pair. This can be treated phenomenologically using four parameters, the transverse and longitudinal relaxation times of each radical, as described in Sect. 3.4. Here we may make two simplifying assumptions: firstly, that the transverse and longitudinal times will be equal ($T_1^{(i)} \simeq T_2^{(i)}$) since the external field is so weak, and secondly that the primary mechanism of relaxation will be modulation of the hyperfine interactions, and so the spin of the electron on the fullerene will not relax ($T_1^{(2)} = T_2^{(2)} = 0$). The consequences of this second assumption will be analysed in more detail later. These approximations reduce the description of relaxation to a single parameter, the relaxation rate of the unpaired electron on the carotenoid radical, $k_R = 1/T_1^{(1)} = 1/T_2^{(1)}$. We shall take this to be a free parameter to fit our simulations to the experimental data, since it is very challenging to determine theoretically or from experiments.

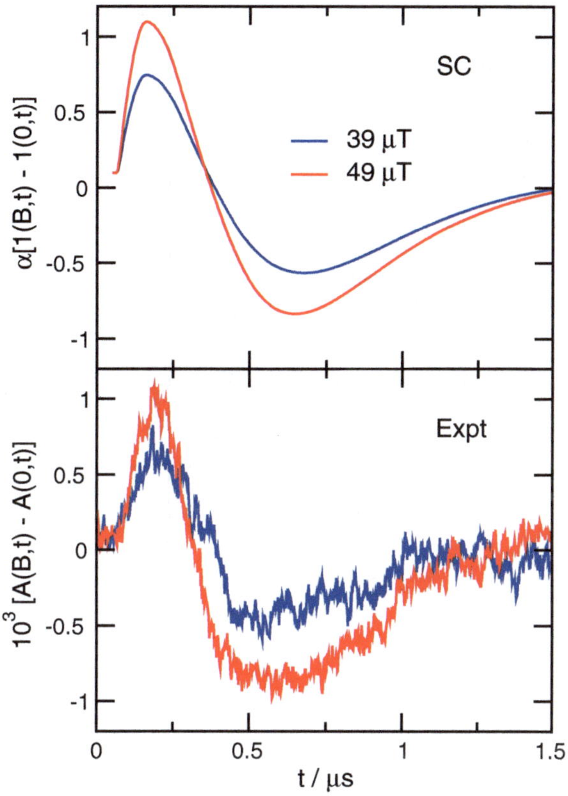

Fig. 5.4 Upper panel: Computed SC magnetic field effects on the total survival probability of the $C^{\bullet+}PF^{\bullet-}$ radical pair at $B = 39$ and 49 μT, multiplied by $\alpha = 821$ to bring the positive peak in the 49 μT signal to 1, and then shifted by 0.1 to mimic the experimental background and delayed by 50 ns to mimic the experimental instrumental delay. Lower panel: Changes in the transient absorption signal of the carotenoid radical in $C^{\bullet+}PF^{\bullet-}$ at 113 K caused by these applied magnetic fields; experimental data from Ref. [2]

5.2.2 Results and Discussion

With these values established, we can compare the results of our semiclassical theory to the experimental data from Ref. [2]. This comparison is shown in Fig. 5.4, with the optimised carotenoid relaxation rate $k_R = 2.3 \times 10^7\ s^{-1}$. With this single parameter, both the timescale and the relative intensity of the positive and negative peaks of the transient absorption curve are reproduced quantitatively at both magnetic field strengths by the SC theory. Furthermore, the absolute magnitude of the magnetic field effect is also reproduced reasonably well: at 0.70 μs, the experimentally measured difference in absorption is approximately 1.5% of the absorption in the absence of a field [2], and the percentage magnetic field effect in the calculations at 0.65 μs (which allows for a ~50 ns instrumental delay) is also 1.5%.

The value of k_R required to fit the experimental data appears to be reasonable. It is five times larger than that obtained by Maeda et al. in a separate experiment, measuring the magnetic field effect on the same triad at 1.28 mT [50]. Given the huge difference in field strength between that experiment and the one discussed here, and the relatively crude model they used to obtain the relaxation rate, we consider

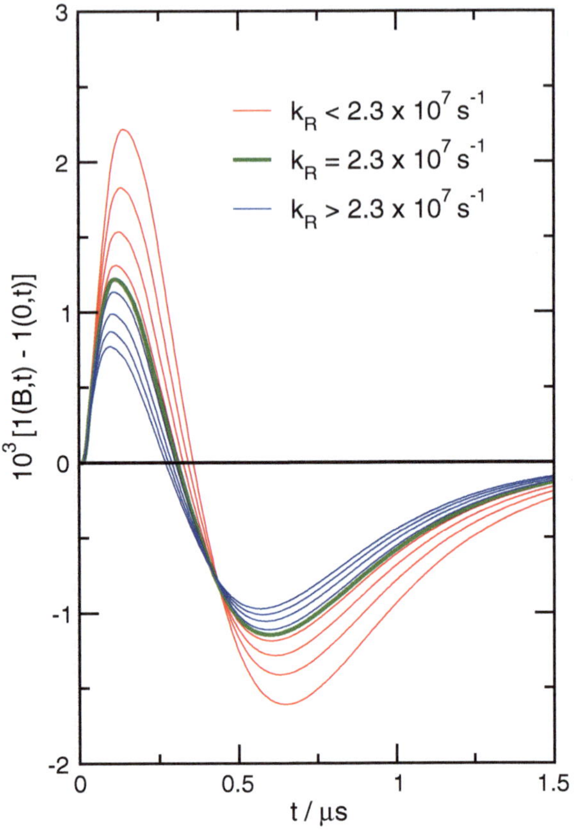

Fig. 5.5 Dependence of the simulated magnetic field effect in the survival probability of $C^{\bullet+}PF^{\bullet-}$ at 49 μT on the relaxation rate k_R of the electron spin on the carotenoid radical. Red curves: $k_R = 1.6 \times 10^7\ s^{-1}$ to $2.2 \times 10^7\ s^{-1}$ in steps of $2 \times 10^6\ s^{-1}$. Green curve: $k_R = 2.3 \times 10^7\ s^{-1}$, which gives the best fit to the experimental data. Blue curves: $k_R = 2.4 \times 10^7\ s^{-1}$ to $3.0 \times 10^7\ s^{-1}$ in steps of $2 \times 10^6\ s^{-1}$

our relaxation rate to be consistent with the value they found. Furthermore, our calculations do not show a particularly strong dependence on the value of k_R chosen. The variation of the magnetic field effect at 49 μT with the relaxation rate is shown in Fig. 5.5. The magnitude and timescale of the magnetic field effect both decrease as the relaxation rate increases, as one might expect. The behaviour is monotonic, making it easy to precisely identify the optimal value of k_R. However, once normalised, the fit is not that sensitive to the precise value of k_R chosen – while $k_R = 2.3 \times 10^7\ s^{-1}$ is definitely the best fit, none of the curves shown in Fig. 5.5 show terrible agreement with the experimental data.

The biphasic form of the magnetic field effect on the transient absorption signal is intriguing. The initial increase in the survival probability is easy to understand – it is a consequence of the low field effect discussed in Sect. 1.3.2 [55, 56]. Since $k_S \gg k_T$, the larger the proportion of radical pairs in the triplet state the larger the survival probability will be. The application of a magnetic field breaks the symmetry of the zero-field triplet eigenstates, allowing more pathways for singlet-triplet interconversion. Since the radical pair is formed predominantly in the singlet state,

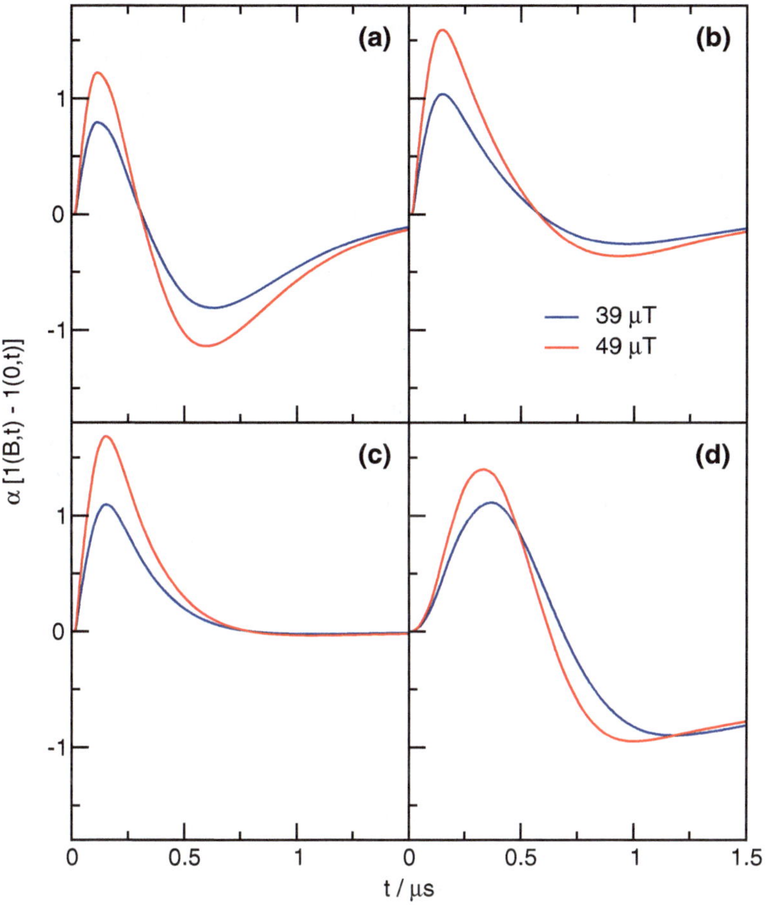

Fig. 5.6 The effect of (artificially) moving the electron spin relaxation from the carotenoid radical (1) to the fullerene radical (2) in the $C^{\bullet+}PF^{\bullet-}$ radical pair, compared to no relaxation at all. **a** Relaxation of the carotenoid electron spin: $k_R^{(1)} = 2.3 \times 10^7\ s^{-1}$ and $k_R^{(2)} = 0$. **b** Relaxation of both electron spins: $k_R^{(1)} = 1.15 \times 10^7\ s^{-1}$ and $k_R^{(2)} = 1.15 \times 10^7\ s^{-1}$. **c** Relaxation of the fullerene electron spin: $k_R^{(1)} = 0$ and $k_R^{(2)} = 2.3 \times 10^7\ s^{-1}$. (d) No relaxation of either electron spin: $k_R^{(1)} = k_R^{(2)} = 0$. In panels **a–c**, $\alpha = 1000$; in panel **d**, $\alpha = 50$

this leads to a larger triplet radical pair population, and hence an increased survival probability, provided that the applied field is not so large as to open a significant energy gap between the triplet and singlet states.

It is harder to understand why at later times the survival probability decreases when a magnetic field is applied. It was originally suggested that this was due to relaxation transferring population from the T_+ and T_- states to the the S and T_0 states [2]. However, Fig. 5.6 shows that this explanation is insufficient: biphasic behaviour is still observed in panel (d) where relaxation is completely neglected. Nevertheless, while relaxation is not necessary for the observation of biphasic behaviour, the experimental data cannot be accounted for without it. Furthermore, it appears that the origin of the relaxation is important. For example, if relaxation occurs (unphysically) on the fullerene radical rather than on the carotenoid radical, as in panel (c), the biphasic behaviour disappears. Including relaxation on both radicals in panel (b) also severely reduces the magnitude of the negative peak. Figure 5.6 also justifies our earlier assumption that the primary relaxation mechanism in the triad is the modulation of hyperfine interactions and so the spin of the unpaired electron on the fullerene radical does not relax, since this is only way to reproduce the experimental data.

The equation of motion of $\mathbf{T}_{12}$, the classical tensor corresponding to the two-electron spin operators, is identical in each of scenarios (a)–(c), precluding any simple population-based interpretation of the effect of relaxation. The equations of motion of the individual electrons are of course different in each case, which gives rise to the different behaviour in Fig. 5.6 through indirect coupling to $\mathbf{T}_{12}$ through the motion of the nuclear spins. Incidentally, this implies that SW theory would be insufficient to reproduce this biphasic signal accurately. All that can be said about the biphasic response for now, then, is that it is the result of a delicate balance between the asymmetry of the recombination rates and the rate and source of relaxation in the radical pair. We will return briefly to this topic in Sect. 7.2.1.

5.3 Anisotropy

Having used the semiclassical theory to reproduce the response of a prototype magnetoreceptor to a magnetic field, we will now apply it to the cryptochrome radical pair proposed as the receptor in European robins. In particular, we will attempt to calculate the anisotropy in the singlet yield of the radical pair, defined as

$$\alpha(\mathbf{B}) = \Phi_S(\mathbf{B}) - \langle \Phi_S(B) \rangle, \tag{5.12}$$

where $\langle \Phi_S(B) \rangle$ is the average of the singlet yield over all direction of the applied field $\mathbf{B} = B(\sin\theta\cos\phi, \sin\theta\sin\phi, \cos\theta)$.

It is convenient to express the anisotropy as a linear combination of spherical harmonics,

$$Y_l^m(\theta, \phi) = N_l^m e^{im\phi} P_l^m(\cos\theta) \tag{5.13}$$

where l can take any non-negative integer, m may take any integer value in the range $-l \leq m \leq l$, N is a normalisation constant, and $P_l^m(x)$ is the associated Legendre polynomial,

$$P_l^m(x) = \frac{(-1)^m \left(1 - x^2\right)^{m/2}}{2^l\, l!} \frac{\mathrm{d}^{l+m}}{\mathrm{d}x^{l+m}} \left(x^2 - 1\right)^l. \tag{5.14}$$

Since $\Phi_S(\mathbf{B}) = \Phi_S(-\mathbf{B})$, the singlet yield may be expressed as function of just the even spherical harmonics,

$$\Phi_S(B, \theta, \phi) \simeq \sum_{l=0,2,4,\ldots}^{L} \sum_{m=-\min(M,l)}^{\min(M,l)} c_{lm}(B) Y_l^m(\theta, \phi), \tag{5.15}$$

where L and M control the accuracy of the approximation, and $M \leq L$. The orthonormality of the spherical harmonics gives

$$c_{lm}(B) = \int_0^{2\pi} d\phi \int_0^{\pi} \sin\theta \, d\theta \, Y_l^m(\theta, \phi)^* \Phi_S(B, \theta, \phi), \tag{5.16}$$

and since $Y_{00}(\theta, \phi)$ is spherically symmetric the anisotropy is

$$\alpha(B, \theta, \phi) = \sum_{l=2,4,\ldots}^{L} \sum_{m=-\min(M,l)}^{\min(M,l)} c_{lm}(B) Y_l^m(\theta, \phi). \tag{5.17}$$

Assuming $\Phi_S(B, \theta, \phi)$ may be represented exactly by the expansion in Eq. 5.15, the integrals in Eq. 5.16 can be evaluated using a $2M + 1$ point quadrature rule in ϕ with weights $w_k = \Delta\phi = 2\pi/(2M + 1)$ and nodes $\phi_k = k\Delta\phi$, and an $L + 2$ point Gauss-Legendre quadrature rule in $\cos\theta$. Since the singlet yield is symmetric with respect to inversion, one can restrict the $L + 2$ point rule to its $(L + 2)/2$ nodes in the upper hemisphere, and multiply the corresponding weights by 2. The quadrature grid therefore contains a total of $(2M + 1)(L + 2)/2$ magnetic field directions.

5.3.1 *Simplified Cryptochrome Models*

We began by calculating the anisotropy in the singlet yield of three simplified cryptochrome models with three, seven and eleven nuclear spins in each radical at a field strength of $B = 1$ mT, both semiclassically and quantum mechanically. In each model, the nuclei whose hyperfine coupling tensors have the largest traces were retained. These anisotropic hyperfine tensors were calculated using unrestricted B3LYP density functional theory [51, 52] with the EPR-II basis set [53], relative to the axis system defined in Fig. 5.2. The hyperfine tensors of the tryptophan were then rotated relative to those of the flavin to bring them into the relative alignment of the DmCry crystal structure in Fig. 5.1. The resulting tensors are listed in Appendix E.

We took the recombination rate constant of both the singlet and triplet radical pairs to be $k = 10^6\ \mathrm{s}^{-1}$, since this is slow enough to allow for significant competition between the singlet and triplet reaction pathways [1, 13, 32, 40], and is similar to the recombination rates found experimentally for [$\mathrm{FAD}^{\bullet-}$ $\mathrm{Trp}_C^{\bullet+}$] radical pairs in related cryptochromes [31, 34]. The singlet yield is defined in Eq. (3.16), and the singlet probability is given within the semiclassical theory by Eq. (3.14) with $\bar{A} = \bar{P}_S$, and quantum mechanically by Eq. (2.48) with $\hat{A} = \hat{P}_S$. For the semiclassical calculations, the integrals in Eq. (3.14) were evaluated by Monte Carlo integration, with one million samples of the orientations of each initial spin vector. The resulting statistical error in the maximum anisotropy of the singlet yield was found to be less than 1%. Both the classical and quantum spin dynamics were run for 11.5 μs. Following preliminary convergence tests, we settled on a quadrature grid of $L = 30$ and $M = 10$, giving a grid of 336 field directions at which the singlet yield was evaluated in the calculations reported below. Increasing M to L (= 30) was found to give results that were indistinguishable to graphical accuracy, at the expense of a significantly larger grid of field directions.

The results of these preliminary calculations are shown in Fig. 5.7, which compares the quantum mechanical (QM) and semiclassical (SC) anisotropies of the three simplified cryptochrome models. The axis system is defined in Fig. 5.2, with the flavin ring system in the xy plane. The QM results are shown on the left and the SC results on the right, for radical pairs with a total of with 6 (top), 14 (middle) and 22 (bottom) nuclear spins. The blue and red regions indicate orientations of **B** for which the anisotropy is positive and negative, respectively. The distance between the origin and the surface of the plot gives the magnitude of the anisotropy when **B** is in that direction. The axes of all six plots have the same scale.

There is a significant difference between the QM and SC calculations of the anisotropy in the singlet yield of the 6-spin cryptochrome model in panels (a) and (b) of Fig. 5.7. The fact that the sharp features in the QM anisotropy are completely absent in the SC anisotropy implies that they are the result of a quantum mechanical interference effect which is not captured by the SC theory. On moving from 6 to 14 to 22 nuclear spins in the radical pair, however, the spikes in the QM anisotropy first shrink and then disappear: the interference effects are removed by increasing the number of nuclear spins. This is consistent with the observations we made when we first introduced the SC theory in Sect. 3.1. The quantum mechanical interference effects in the electron spin dynamics are rapidly quenched as the number of nuclear spins increases, so there is no need to use a full QM calculation which includes them. All that survives by the time there are 22 nuclear spins in the radical pair is the average semiclassical precessional behaviour, which a comparison of panels (e) and (f) in Fig. 5.7 shows the SC theory captures extremely well.

Unfortunately, investigating the singlet yield anisotropy at lower field strengths is more difficult in the SC theory, because of the statistical error involved in evaluating the the integrals in Eq. (3.14) by a Monte Carlo sampling method. In an Earth-strength magnetic field (~50 μT), the difference between the maximum and minimum anisotropies is only a small fraction of the spherically averaged singlet yield $\langle \Phi_S(B) \rangle$, and therefore the ratio of signal-to-statistical noise in the relative anisotropy is very low. Many more Monte Carlo points have to be sampled to overcome this, and since the precise calculation of the anisotropy requires a dense grid of magnetic field directions, the calculation becomes unreasonably expensive. Therefore, at very low fields we cannot validate the SC anisotropy against exact QM results in the same way as we have in Fig. 5.7.

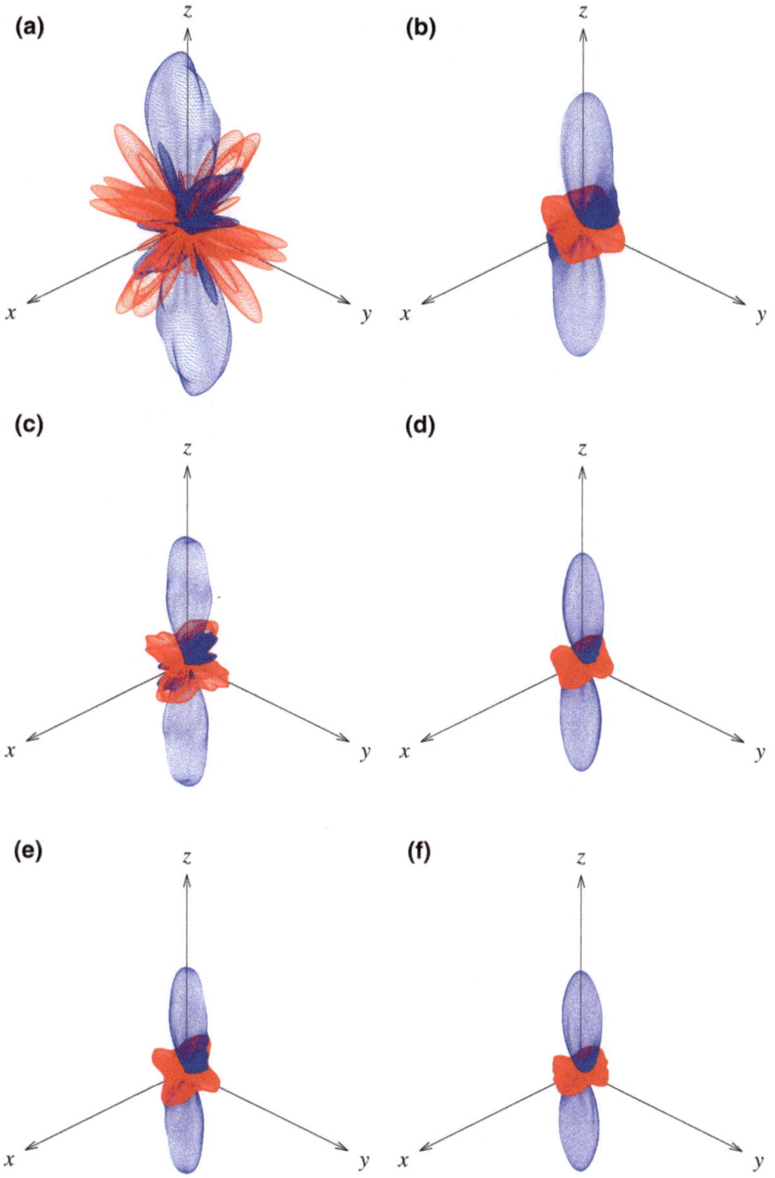

Fig. 5.7 Anisotropies in the singlet yields of three simplified cryptochrome models. Blue indicates a positive value of α, red a negative value. The QM results are shown on the left and the SC results on the right, for [$FAD^{\bullet-}$ $Trp_C^{\bullet+}$] radical pairs with 6 (top), 14 (middle) and 22 (bottom) nuclear spins. The axis scales are the same in all six plots, with the axes oriented relative to the flavin radical as illustrated in Fig. 5.2. The maximum and minimum anisotropies, $\alpha_{\max}$ and $\alpha_{\min}$, are: **a** 0.0169 and −0.0124; **b** 0.0136 and −0.0042; **c** 0.0126 and −0.0046; **d** 0.0110 and −0.0031; **e** 0.0104 and −0.0038; and **f** 0.0101 and −0.0029

Inspection of Fig. 5.7 suggests a simple solution to this problem: in this particular radical pair the direction of maximum anisotropy is close to the z axis and that of minimum anisotropy is close to the x axis. These axes are defined relative to the orientation of the $FAD^{\bullet -}$ radical, which is independent of the magnetic field strength, so we would expect the same to be true for lower applied magnetic fields. This suggests that subtracting the singlet yield obtained when the applied magnetic field is along the x axis from the singlet yield calculated with the field aligned with the z axis will provide a reasonable estimate of the magnitude of the anisotropy,

$$\Delta\alpha \simeq \alpha_z - \alpha_x = \Phi_S^z - \Phi_S^x. \tag{5.18}$$

Despite the crudeness of this approximation, we have found that $\Delta\alpha$ is within 15% of the true magnitude of the anisotropy at magnetic field strengths of 1, 0.75, 0.5, and 0.25 mT, with no systematic degradation in the accuracy of the approximation as the field strength is reduced.

With this simplified approach, we were able to use the SC theory to estimate the magnitude of the anisotropy in the singlet yield of the 22-spin model of the cryptochrome radical pair down to geomagnetic field strengths. In Fig. 5.8 we compare its performance against exact calculations of $\Delta\alpha$. The good agreement between the SC theory and QM simulations shows that the semiclassical approach can at the very least be relied upon to provide a reasonable estimate of magnitude of the anisotropy in the singlet yield at an Earth-strength magnetic field.

At 50 μT, the anisotropy in the singlet yield of the 22-spin model of the cryptochrome radical pair is $\Delta\alpha = 3.1 \times 10^{-4}$. This is about a quarter of the value quoted by Lee et al. in Ref. [38], where they calculated the anisotropy of a 14-spin

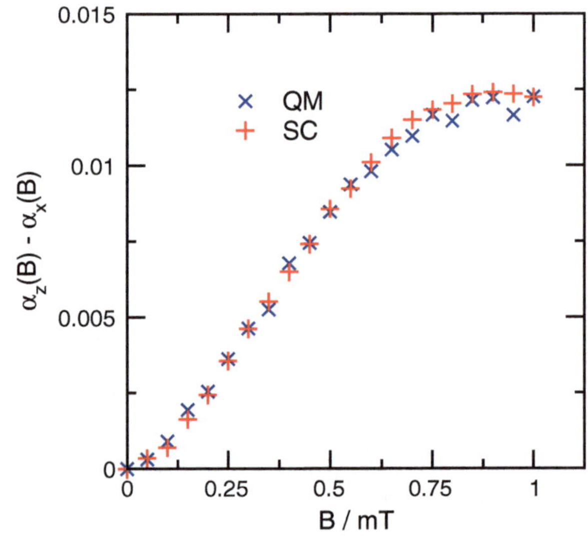

Fig. 5.8 Magnetic field dependence of the QM and SC $\Delta\alpha \simeq \alpha_z - \alpha_x$ for the 22-spin cryptochrome model. In all cases the statistical errors in the SC calculation are smaller than the sizes of the plotted points

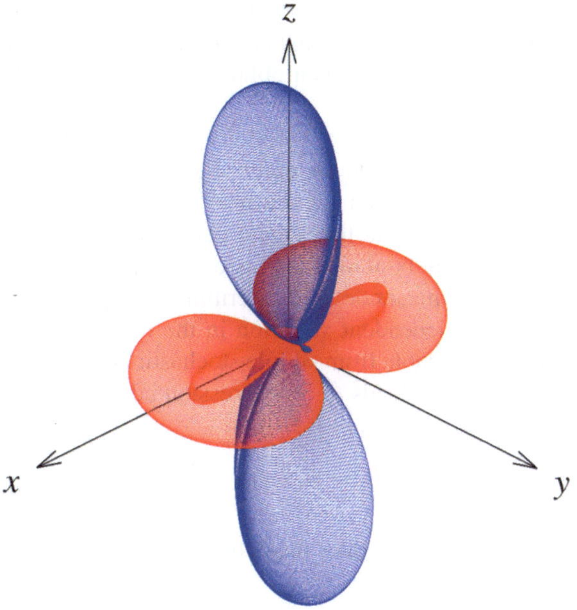

Fig. 5.9 The anisotropy in the singlet yield of the full cryptochrome system, including dipolar and exchange coupling between the electron spins, in a 50 μT magnetic field; $\alpha_{\text{max}} = 3.5 \times 10^{-5}$ and $\alpha_{\text{min}} = -2.7 \times 10^{-5}$

cryptochrome model. It is striking that the inclusion of even very small hyperfine interactions has a significant effect on the properties of the anisotropy. However, this calculation still neglects five hyperfine interactions and, more importantly, the exchange and dipolar coupling between the electron spins. These have not been included until now because doing so makes the QM calculation intractable, since the techniques described in Sect. 2.6 can no longer be used. However, they may be straightforwardly included in the SC theory with very little computational cost, as described in Appendix C.

5.3.2 The Full Cryptochrome Radical Pair

Figure 5.9 shows the anisotropy in the singlet yield of the full cryptochrome radical pair at 50 μT using spherical harmonics of order up to 4 (using $L = M = 4$ in Eq. (5.15) and a grid of 27 magnetic field directions), averaging over 100 million trajectories for each field direction. The dipolar coupling constant was given by the point dipole approximation as $D = -0.4\,\text{mT}$, and we took the exchange coupling constant to be $J = -2/3 D = 0.27\,\text{mT}$, close to that determined experimentally for a related cryptochrome radical pair [33]. The resulting anisotropy has a similar shape to the anisotropy in a 1 mT field in Fig. 5.7, and is even smaller in magnitude than that of the 22-spin model of the radical pair: $\alpha_{\text{max}} - \alpha_{\text{min}} = (6.2 \pm 1.1) \times 10^{-5}$.[3]

[3]The cruder estimate for the anisotropy used earlier is $\alpha_z - \alpha_x = (4.5 \pm 1.2) \times 10^{-5}$.

This represents a $(0.022 \pm 0.004)\%$ change in the total singlet yield, far smaller than all previous estimates of the anisotropy of the cryptochrome radical pair at this field strength, which have been obtained from simplified models and have ranged from 0.5 to 10% [13, 38].

This response to the direction of the applied magnetic field can be used to provide a rough estimate the number of receptors required to achieve a signal to noise ratio of 1, for a given angular resolution [39, 40, 57]. These estimates are based on the assumption that the radical pair reaction leads to the production of signalling molecules that bind to receptors on the optic nerve [1], and therefore that the orientation of the magnetic field determines the quantity of these signalling ligands. This orientation-dependent variation in the number of ligand-receptor complexes must be distinguishable from simple stochastic fluctuations in the concentration of bound receptors. This leads to an expression adapted from Weaver et al. for the required number of receptors [40, 57],

$$R = \frac{4}{(\lambda_\theta \Delta\theta)^2}, \tag{5.19}$$

where $\Delta\theta$ is the angular resolution of the proposed compass in degrees, and

$$\lambda_\theta = \frac{1}{90^\circ} \frac{\alpha_z - \alpha_x}{\langle \Phi \rangle}. \tag{5.20}$$

Using this expression, we determine that for cryptochrome to provide a geomagnetic compass accurate to 1°, between 5×10^{11} and 10^{12} receptors would be required in the retina of the bird. This is significantly higher than all previous estimates, which have ranged from 10^5 to 10^9 [58]. Furthermore, in this estimate we have assumed a perfectly ordered arrangement of cryptochrome proteins; if there is some disorder in their orientation, more receptors would be required to compensate for this [59].

While there are (necessarily) a great many simplifications in an argument such as this, the calculation of a ~0.02% change in the singlet yield of the cryptochrome radical pair upon changing the orientation of the applied magnetic field is hardly compelling evidence for the cryptochrome radical pair described acting as a biological magnetoreceptor. Two suggestions have been made which would increase the anisotropic sensitivity of the radical pair, making it more plausible as a magnetoreceptor. The first is the possibility of an alternative counter-radical for $FAD^{\bullet-}$, ideally with few or no hyperfine interactions, as discussed in Sect. 5.1.2. While an alternative radical pair of this type would certainly have a greater anisotropic sensitivity [38], and be consistent with some of the RF experiments [2, 23], it is not currently obvious what the identity of the new counter-radical might be. The second suggestion is the possibility that the $[FAD^{\bullet-}\ Trp_C^{\bullet+}]$ radical pair is much longer-lived than previously thought, which is implied by some of the RF experiments on European robins [13, 21]. The consequences of reducing the recombination rate of this radical pair are discussed briefly in the following section.

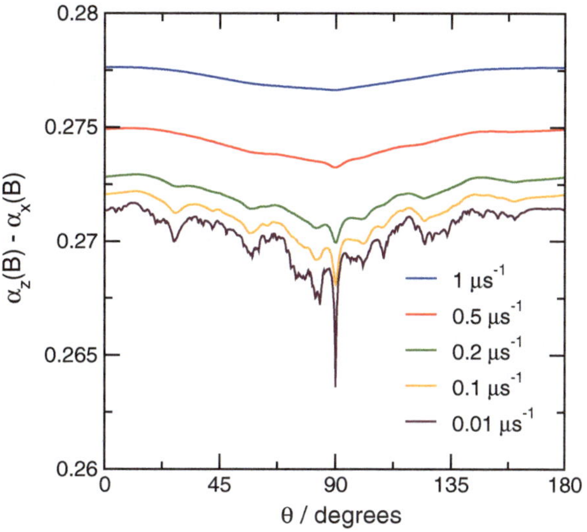

Fig. 5.10 The variation of the singlet yield of the 14-spin cryptochrome model as a function of the orientation of the $50\,\mu$T applied field for a series of decreasing recombination rates. θ is measured relative to the z axis in the xz plane. Data is reproduced with permission from Ref. [60]

5.3.3 A Compass Needle?

A recent study by Hiscock et al. investigated the effect of increasing the lifetime of the cryptochrome radical pair [60]. They performed QM calculations of the singlet yield of the 14- and 22-spin models of the [$FAD^{\bullet-}$ $Trp_C^{\bullet+}$] radical pair with a series of decreasing symmetric recombination rates. The effect of rotating the applied $50\,\mu$T field through an angle θ from the z axis in the xz plane are shown in Fig. 5.10, which shows data taken from Ref. [60]. At short lifetimes, the singlet yield shows only a weak, smooth dependence on θ, consistent with the results in Sects. 5.3.1 and 5.3.2, as well as earlier work [38]. However, as the lifetime of the radical pair increases, much more structure emerges, most notably a strong "spike" in the singlet yield when $\theta = 90°$. It has been suggested that this spike could act as a compass needle for birds, a mechanism which would allow them to orient themselves precisely in the Earth's magnetic field [27, 49, 60].

This spike arises from an avoided crossing of the quantum mechanical energy levels arising from the N5 and N10 hyperfine tensors [60]. There are several conditions which must be met for the spike to be observed. The first, as demonstrated in Fig. 5.10, is a small recombination rate constant k. In addition, the A_{xx} and A_{yy} components of the N5 and N10 hyperfine tensors must be small relative to A_{zz}, but non-zero, otherwise the avoided crossings become level crossings. Finally, if the counter-radical to $FAD^{\bullet-}$ has no anisotropic hyperfine interactions, no spike is observed for any value of the recombination rate.

The lifetime of the radical pair required for the spike in the anisotropy to be observed appears at first sight to be implausibly long. While the recombination rate of the radical pair could well be significantly smaller than previously assumed [1, 13], it seems less likely that the relaxation rate and corresponding decoherence of the radical pair could be slow enough to allow the spike to emerge [26]. In order for relaxation to be sufficiently slow for the spike to be observed, the librations of the tryptophan radical in particular would have to be extremely fast [27]. Furthermore, the calculations in Ref. [60] neglect electron spin coupling; it would be interesting to know if the avoided crossing, and hence the spike, is still observed when coupling is included. Unfortunately, at present the QM calculation required to investigate this is not feasible.

Nevertheless, the natural emergence of this spike in the anisotropy of the singlet yield of the cryptochrome radical pair which was originally proposed as the magnetoreceptor in the avian compass is fairly compelling. This mechanism of magnetoreception would provide a satisfying account of the precision of the compass [7, 8], and could possibly explain its response to RF fields [19–22]. Finally, it should be noted that this is a genuinely quantum mechanical effect: no spike is observed in the equivalent semiclassical calculations. If this were indeed the mechanism of magnetoreception, the needle of the avian compass would be a true example of quantum biology [61–63].

5.4 Conclusion

In this chapter, we began by considering the experiments which have been performed to investigate the compass sense of migratory birds. These provide some evidence for the radical pair mechanism of magnetoreception, and indicate some properties which a candidate radical pair must possess in order to function as a suitable magnetoreceptor. We then introduced the cryptochrome-based radical pair thought to provide directional information to European robins.

Using the semiclassical theory introduced in Sect. 3.1, we were able to reproduce the experiments of Maeda et al. which demonstrated an Earth-strength magnetic field effect on a radical pair reaction for the first time [2]. In doing so, we were able to shed light on the biphasic nature of the MFE they found. This results from a delicate balance between the asymmetry in the singlet and triplet recombination rates, and the rate and source of electron spin relaxation.

We then calculated the anisotropy of the cryptochrome radical pair. We began by validating our semiclassical results against exact quantum mechanical calculations, and then used the SC model to investigate a more detailed model of the cryptochrome radical pair than it is possible to simulate with a fully quantum mechanical method. We found that once all of the hyperfine interactions and the electron spin coupling are accounted for, only a very small anisotropy remains. It is hard to see how this weak effect could translate into useful directional information for migrating birds. We have ended the chapter by briefly discussing a recent development which has suggested that the radical pair could exhibit a much more precise anisotropic response if it were very long-lived [60].

References

1. Ritz, T., Adem, S., & Schulten, K. (2000). A model for photoreceptor-based magnetoreception in birds. *Biophysical Journal*, *78*, 707–718.
2. Maeda, K., et al. (2008). Chemical compass model of avian magnetoreception. *Nature*, *453*, 387–390.
3. Von Middendorf, A. (1859). Die Isepiptesen Rufflands. Mem. Acad. Sci. St Petersbourg VI, Ser. Tome 8, 1–143.
4. Viguier, C. Le. (1882). Sens de L'orientation et ses Organes chez les Animaux et chez L'homme. *Revue Philosophique de la France et de l' Étranger*, *14*, 1–36.
5. Wiltschko, W. (1968). Über den Einfluß statischer Magnetfelder auf die Zugorientierung der Rotkehlchen (Erithacus rubecula). *Zeitschrift für Tierpsychologie*, *25*, 537–558.
6. Wiltschko, W., & Wiltschko, R. (1996). Magnetic orientation in birds. *Journal of Experimental Biology*, *199*, 29–38.
7. Åkesson, S., Morin, J., Muheim, R., & Ottosson, U. (2001). Avian orientation at steep angles of inclination: experiments with migratory white-crowned sparrows at the magnetic North Pole. *Proceedings of the Royal Society B: Biological Sciences*, *268*, 1907–1913.
8. Lefeldt, N., Dreyer, D., Schneider, N.-L., Steenken, F., & Mouritsen, H. (2015). Migratory blackcaps tested in Emlen funnels can orient at 85 degrees but not at 88 degrees magnetic inclination. *Journal of Experimental Biology*, *218*, 206–211.
9. Wiltschko, W., & Wiltschko, R. (1972). Magnetic compass of European robins. *Science*, *176*, 62–64.
10. Wiltschko, W., & Wiltschko, R. (1992). Migratory orientation: Magnetic compass orientation of garden warblers (Sylvia borin) after a simulated crossing of the magnetic equator. *Ethology*, *91*, 70–74.
11. Sakurai, J. J. (1986). *Modern quantum mechanics*. London: Pearson.
12. Wiltschko, W. (1978). *Animal migration, navigation, and homing* (pp. 302–310). Berlin: Springer.
13. Rodgers, C. T., & Hore, P. J. (2009). Chemical magnetoreception in birds: the radical pair mechanism. *Proceedings of the National Academy of Sciences of the United States of America*, *106*, 353–360.
14. Wiltschko, W., & Wiltschko, R. (1995). Migratory orientation of European Robins is affected by the wavelength of light as well as by a magnetic pulse. *Journal of Comparative Physiology A*, *177*, 363–369.
15. Johnsen, S., & Lohmann, K. J. (2005). The Physics and Neurobiology of Magnetoreception. *Nature Reviews Neuroscience*, *6*, 703–712.
16. Timmel, C. R., & Hore, P. J. (1996). Oscillating magnetic field effects on the yields of radical pair reactions. *Chemical Physics Letters*, *257*, 401–408.
17. Henbest, K. B., Kukura, P., Rodgers, C. T., Hore, P. J., & Timmel, C. R. (2004). Radio frequency magnetic field effects on a radical recombination reaction: A diagnostic test for the radical pair mechanism. *Journal of the American Chemical Society*, *126*, 8102–8103.
18. Woodward, J. R., Timmel, C. R., McLauchlan, K. A., & Hore, P. J. (2001). Radio frequency magnetic field effects on electron-hole recombination. *Physical Review Letters*, *87*, 077602.
19. Ritz, T., Thalau, P., Phillips, J. B., & Wiltschko, R. (2004). Resonance effects indicate a radical-pair mechanism for avian magnetic compass. *Nature*, *429*, 177–180.
20. Thalau, P., Ritz, T., & Stapput, K. (2005). Magnetic compass orientation of migratory birds in the presence of a 1.315 MHz oscillating field. *Naturwissenschaften*, *92*, 86–90.
21. Engels, S., et al. (2014). Anthropogenic electromagnetic noise disrupts magnetic compass orientation in a migratory bird. *Nature*, *509*, 353–356.
22. Wiltschko, R. (2015). Magnetoreception in birds: The effect of radio-frequency fields. *Journal of the Royal Society Interface*, *12*, 20141103.
23. Rodgers, C. T. (2007). *Magnetic Field Effects in Chemical Systems*, Ph.D. thesis, University of Oxford.

24. Schwarze, S. (2016). Weak broadband electromagnetic fields are more disruptive to magnetic compass orientation in a night-migratory songbird (erithacus rubecula) than strong narrowband fields. *Frontiers in Behavioral Neuroscience*, *10*, 55.
25. Lau, J. C. S., Wagner-Rundell, N., Rodgers, C. T., Green, N. J. B., & Hore, P. J. (2009). Effects of disorder and motion in a radical pair magnetoreceptor. *Journal of the Royal Society Interface*, *7*, 257–264.
26. Kattnig, D. R., Solov'yov, I. A., & Hore, P. J. (2016). Electron spin relaxation in cryptochrome-based magnetoreception. *Physical Chemistry Chemical Physics*, *18*, 12443–12456.
27. Worster, S., Kattnig, D. R., & Hore, P. J. (2016). Spin relaxation of radicals in cryptochrome and its role in avian magnetoreception. *Journal of Chemical Physics*, *145*, 035104.
28. Zoltowski, B. D., et al. (2011). Structure of full-length Drosophila cryptochrome. *Nature*, *480*, 396–399.
29. Levy, C., et al. (2013). Updated structure of Drosophila cryptochrome. *Nature*, *495*, E3–E4.
30. Brautigam, C. A., et al. (2004). Structure of the photolyase-like domain of cryptochrome 1 from Arabidopsis thaliana. *Proceedings of the National Academy of Sciences of the United States of America*, *101*, 12142–12147.
31. Aubert, C., Vos, M. H., Mathis, P., & Brettel, K. (2000). Intraprotein radical transfer during photoactivation of DNA photolyase. *Nature*, *405*, 586–590.
32. Solov'yov, I. A., Chandler, D. E., & Schulten, K. (2007). Magnetic field effects in arabidopsis thaliana cryptochrome-1. *Biophysical Journal*, *92*, 2711–2726.
33. Biskup, T., et al. (2009). Direct observation of a photoinduced radical pair in a cryptochrome blue-light photoreceptor. *Angewandte Chemie - International Edition*, *48*, 404–407.
34. Maeda, K., et al. (2012). Magnetically sensitive light-induced reactions in cryptochrome are consistent with its proposed role as a magnetoreceptor. *Proceedings of the National Academy of Sciences of the United States of America*, *109*, 4774–4779.
35. Dodson, C. A., Hore, P. J., & Wallace, M. I. (2013). A radical sense of direction: signaling and mechanism in cryptochrome magnetoreception. *Trends in Biochemical Sciences*, *38*, 435–446.
36. Nießner, C. (2011). Avian ultraviolet/violet cones identified as probable magnetoreceptors. *PLOS ONE*, *6*, e20091.
37. Mouritsen, H. (2013). *Neurosciences - from molecule to behavior: a university textbook* (pp. 427–443). Berlin: Springer.
38. Lee, A. A., et al. (2014). Alternative radical pairs for cryptochrome-based magnetoreception. *Journal of the Royal Society Interface*, *11*, 20131063.
39. Efimova, O., & Hore, P. J. (2008). Role of exchange and dipolar interactions in the radical pair model of the avian magnetic compass. *Biophysical Journal*, *94*, 1565–1574.
40. Cintolesi, F., Ritz, T., Kay, C. W. M., Timmel, C. R., & Hore, P. J. (2003). Anisotropic recombination of an immobilized photoinduced radical pair in a 50-μT magnetic field: A model avian photomagnetoreceptor. *Chemical Physics*, *294*, 385–399.
41. Solov'yov, I. A., & Schulten, K. (2009). Magnetoreception through cryptochrome may involve superoxide. *Biophysical Journal*, *96*, 4804–4813.
42. Müller, P., & Ahmad, M. (2011). Light-activated cryptochrome reacts with molecular oxygen to form a flavin-superoxide radical pair consistent with magnetoreception. *Journal of Biological Chemistry*, *286*, 21033–21040.
43. Karogodina, T. Y., Dranov, I. G., Sergeeva, S. V., Stass, D. V., & Steiner, U. E. (2011). Kinetic magnetic-field effect involving the small biologically relevant inorganic radicals NO and $O2^{\bullet-}$. *ChemPhysChem*, *12*, 1714–1728.
44. Kalmijn, A. J., & Blakemore, R. P. (1978). *Animal migration, navigation, and homing* (pp. 354–355). Berlin: Springer.
45. Kirschvink, J. L., Walker, M. M., & Diebel, C. E. (2001). Magnetite-based magnetoreception. *Current Opinion in Neurobiology*, *11*, 462–467.
46. Frankel, R. B., & Blakemore, R. P. (1989). Magnetite and magnetotaxis in microorganisms. *Bioelectromagnetics*, *10*, 223–237.
47. Grissom, C. B. (1995). Magnetic field effects in biology: A survey of possible mechanisms with emphasis on radical-pair recombination. *Chemical Reviews*, *95*, 3–24.

48. Kirschvink, J. L., Winklhofer, M., & Walker, M. M. (2010). Biophysics of magnetic orientation: strengthening the interface between theory and experimental design. *Journal of the Royal Society Interface*, *7*, S179–S191.
49. Hore, P. J., & Mouritsen, H. (2016). The radical-pair mechanism of magnetoreception. *Annual review of Biophysics*, *45*, 299.
50. Maeda, K., et al. (2011). Spin-selective recombination kinetics of a model chemical magnetoreceptor. *Chemical Communications*, *47*, 6563–6565.
51. Lee, C., Yang, W., & Parr, R. G. (1988). Development of the Colle-Salvetti correlation-energy formula into a functional of the electron density. *Physical Review B*, *37*, 785–789.
52. Becke, A. D. (1993). A new mixing of Hartree-Fock and local density-functional theories. *The Journal of Chemical Physics*, *98*, 1372.
53. Barone, V. (1995). *Recent Advances in Density Functional Methods: Part I*. Singapore: World Scientific.
54. Newman, T. (2003). *Exploring the chemical compass: An experimental and theoretical study Part II*, Thesis, University of Oxford.
55. Timmel, C. R., Till, U., Brocklehurst, B., Mclauchlan, K. A., & Hore, P. J. (1998). Effects of weak magnetic fields on free radical recombination reactions. *Molecular Physics*, *95*, 71–89.
56. Till, U., Timmel, C. R., Brocklehurst, B., & Hore, P. J. (1998). The inuence of very small magnetic fields on radical recombination reactions in the limit of slow recombination. *Chemical Physics Letters*, *298*, 7–14.
57. Weaver, J. C., Vaughan, T. E., & Astumian, R. D. (2000). Biological sensing of small field differences bymagnetically sensitive chemical reactions. *Nature*, *405*, 707–709.
58. Mouritsen, H., & Hore, P. J. (2012). The magnetic retina: Light-dependent and trigeminal magnetoreception in migratory birds. *Current Opinion in Neurobiology*, *22*, 343–352.
59. Hill, E., & Ritz, T. (2010). Can disordered radical pair systems provide a basis for a magnetic compass in animals? *Journal of the Royal Society Interface*, *7*(Suppl 2), S265–71.
60. Hiscock, H. G., et al. (2016). The quantum needle of the avian magnetic compass. *Proceedings of the National Academy of Sciences of the United States of America*, *113*, 201600341.
61. Ball, P. (2011). The dawn of quantum biology. *Nature*, *474*, 272–274.
62. Al-Khalili, J. & McFadden, J. (2014). Life on the edge: The coming of ageof quantum biology, Transworld.
63. Kominis, I. K. (2015). The radical-pair mechanism as a paradigm for the emerging science of quantum biology. *Modern Physics Letters B*, *29*, 1530013.

Chapter 6
Magnetoelectroluminescence

Electroluminescence is an important and much studied property of semiconducting films of conjugated organic polymers [1–4], and is the basis of their commercial application in organic light emitting diodes (oLEDs) [5–8]. These have the potential to be more efficient, more easily scalable, and more flexible than their inorganic counterparts [7, 9, 10]. oLEDs are constructed in four layers: a thin film of the semiconducting polymer is sandwiched between an electron-injecting metal cathode and a transparent hole-injecting layer, which is then covered by a transparent anode. Calcium and aluminium are commonly used for the cathode and indium tin oxide for the anode, with poly(3,4-ethylenedioxythiophene)-poly(styrenesulfonate) (PEDOT-PSS) a typical hole-injecting layer [11–15]. In this chapter, we are concerned with the properties of the semiconducting polymer layer which affect the efficiency of electroluminescence.

As a result of the commercial applications of oLEDs, there is significant interest in understanding the mechanism of the electroluminescence of organic semiconductors and the factors which influence it [2–4, 9]. The observation of magnetoelectroluminescence (MEL) – that is, a change in the electroluminescence of a polymer upon application of a magnetic field – has provided evidence for the polaron pair (PP) mechanism of electroluminescence described in Sect. 6.1 [9, 16]. However, the physical interactions which govern the spin dynamics of the polaron pair are less well understood. Recent isotopic substitution experiments have strongly suggested that the hyperfine interactions between the electron spin and the spins of the hydrogen nuclei in the polymer play a crucial role [13, 14], but it remains unclear whether other physical effects are also important.

In this chapter we will describe the polaron pair mechanism of magnetoelectroluminescence, and analyse the connection between MEL and magnetoconductance (MC) in order to define a general relationship between these two properties and the singlet yield of the polaron pair recombination reaction. With this expression, we will reproduce the results of the pioneering deuteration experiments of Nguyen et al.[13, 14] using Schulten–Wolynes (SW) theory, which was introduced in Sect. 3.2,

A. Lewis, *Spin Dynamics in Radical Pairs*, Springer Theses,
https://doi.org/10.1007/978-3-030-00686-0_6

to describe the spin dynamics of the polaron pair. This simple approach is expected to be reasonably reliable, since each polaron contains around $N \sim 100$ nuclear spins. Finally, we will discuss how the singlet-triplet dephasing arising from the modulation of the exchange interaction between the polarons affects magnetoelectroluminescence. We will show that including this phenomenon in our theoretical model, in addition to the hyperfine interactions, resolves some of the discrepancies between theory and experiment.[1]

6.1 The Polaron Pair Mechanism

It is widely believed that electroluminescence in organic polymers occurs through the polaron pair mechanism [9, 14, 16–19]. A polaron is simply an electron (or hole) which is coupled to the nuclear degrees of freedom of the polymer on which it resides [20]. Polarons become localised on a conjugated polymer chain by a distortion of the carbon bond framework. This is either described as "self-localisation", if the localisation results from nuclear displacements induced by the interaction between the electronic and nuclear degrees of freedom [21, 22], or "Anderson localisation", if the localisation is due to conformational disorder in the polymer [23, 24]. We expect the polarons in the semiconducting polymers considered in this chapter to be primarily localised by torsional Anderson disorder, so nuclear motions can be safely neglected [21].

A pair of weakly bound polarons of opposite charges localised on adjacent polymer chains are called a polaron pair [16]. The existence of polaron pairs in films of semiconducting polymers has been shown using photoinduced absorption experiments [25, 26], and reaction yield detected magnetic resonance (RYDMR) and electron spin resonance (ESR) experiments have demonstrated that they may have either singlet or triplet character [27, 28]. A simplified reaction scheme for polaron pairs in oLEDs is given in Fig. 6.1. Positive and negative polarons are injected into and move through a polymer film until they encounter a polaron of opposite charge on an adjacent polymer chain, when they form a loosely bound polaron pair. Since the spins of the polarons are initially uncorrelated, singlet and triplet polaron pairs are formed in a statistical ratio of 1:3. A singlet (or triplet) polaron pair may then form an intra-chain exciton at a rate k_{SE} (k_{TE}), with the electron hopping onto the polymer chain on which the hole resides. Alternatively, the polaron pair could dissociate to reform free polarons at a rate k_{SD} (k_{TD}). If the total decay rates $k_{\mathrm{S}} = k_{\mathrm{SD}} + k_{\mathrm{SE}}$ and $k_{\mathrm{T}} = k_{\mathrm{TD}} + k_{\mathrm{TE}}$ are sufficiently small, significant interconversion between the singlet and triplet states of the polaron pair is possible, due to the hyperfine interactions within each polaron. This description of polaron pairs is almost entirely equivalent

[1] Parts of this chapter have been reproduced with permission from Lawrence, J. E., Lewis, A. M., Manolopoulos, D. E. & Hore, P. J. *Journal of Chemical Physics* **144**, 214109 (2016), https://aip.scitation.org/doi/10.1063/1.4953093

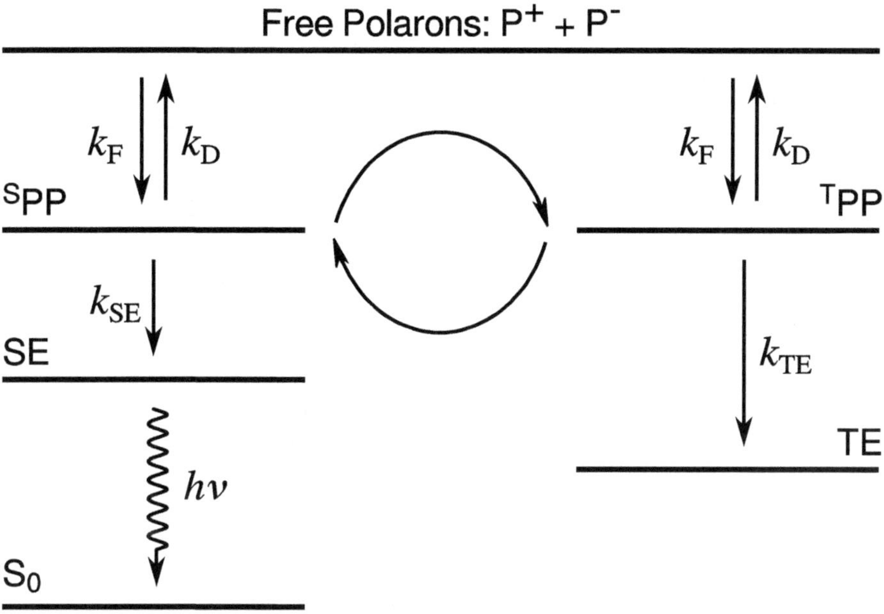

Fig. 6.1 Reaction scheme for polarons. ^{S}PP and ^{T}PP refer to coulombically bound polaron pairs on neighbouring chains in the singlet and triplet state respectively; the curved arrows represent hyperfine-mediated intersystem crossing between the two spin states. SE and TE refer to intra-chain singlet and triplet excitons respectively. Only the singlet exciton state is emissive. The total singlet and triplet decay rates are $k_S = k_{SE} + k_{SD}$ and $k_T = k_{TE} + k_{TD}$

to that of the radical pair mechanism, and so we may apply any of the techniques outlined in Chaps. 2 and 3 to this problem.

Provided that $k_S \neq k_T$, changes in the rate of interconversion will affect the singlet exciton yield. Therefore, since only the singlet exciton is emissive, and the application of a magnetic field alters the rate of interconversion between spin states, the polaron pair mechanism gives rise to magnetoelectroluminescence (MEL),

$$\mathrm{MEL}(B) = \frac{\mathrm{EL}(B)}{\mathrm{EL}(0)} - 1, \tag{6.1}$$

where $\mathrm{EL}(B)$ is the electroluminescence in the presence of a magnetic field of strength B. Indeed, the observation of MEL in organic semiconductors is good evidence for the polaron pair mechanism of electroluminescence [9, 16], because the singlet and triplet intra-chain excitons are well separated in energy by the exchange interaction and so do not easily interconvert [29].

A magnetic field effect in the conductance of organic polymers has also been observed. However, the origin of this effect is not well understood, and is the subject of continuing debate [27, 30–33]. We shall attempt to avoid this issue in this chapter by making as few assumptions as possible about the mechanism of magnetoconductance

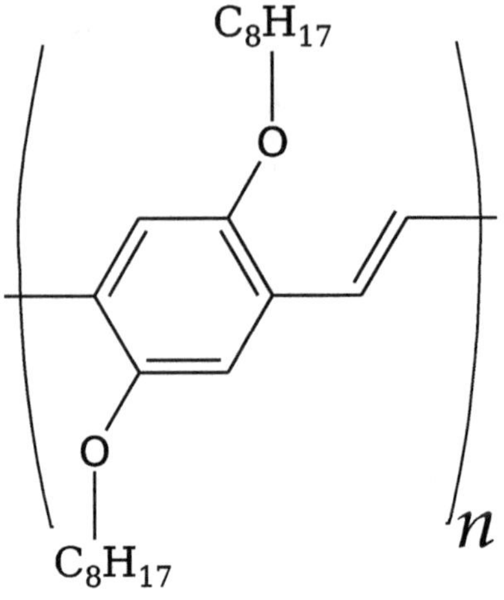

Fig. 6.2 The repeat unit of 2,5-dioctyloxy-paraphenylene vinylene (DOO-PPV)

(MC),

$$\mathrm{MC}(B) = \frac{\mathrm{C}(B)}{\mathrm{C}(0)} - 1, \tag{6.2}$$

where $\mathrm{C}(B)$ is the conductance in the presence of a magnetic field of strength B.

Both the MEL and MC have been measured in a variety of materials. In this work, we will consider poly(2,5-dioctyloxy-paraphenylene vinylene) (DOO-PPV), whose repeat unit is shown in Fig. 6.2. This polymer was chosen because the impact of deuteration on its MEL and MC has been studied [13, 14]. The observation of an isotope effect on MEL and MC has provided evidence for a hyperfine-mediated mechanism of intersystem crossing of the polaron pairs, since D has a different spin quantum number and gyromagnetic ratio to H and hence different hyperfine interactions. The experimentally determined MEL and MC in both undeuterated DOO-PPV (denoted H-DOO-PPV) and deuterated DOO-PPV (denoted D-DOO-PPV) are shown in Fig. 6.3. In order to compare these experimental measurements to theoretical calculations, we must now to relate the MEL and MC to the singlet yield of the polaron pair recombination reaction.

6.2 The Relationship Between MEL and MC

The electroluminescence of a semiconducting polymer at a given magnetic field strength depends not only on the singlet exciton yield, but also on the number density of polaron pairs present in the device, n_{PP}. Previous analyses of MEL have all been undertaken with an implicit assumption that the number of polaron pairs in the device is independent of the applied magnetic field [13, 16, 17]. This leads to a very

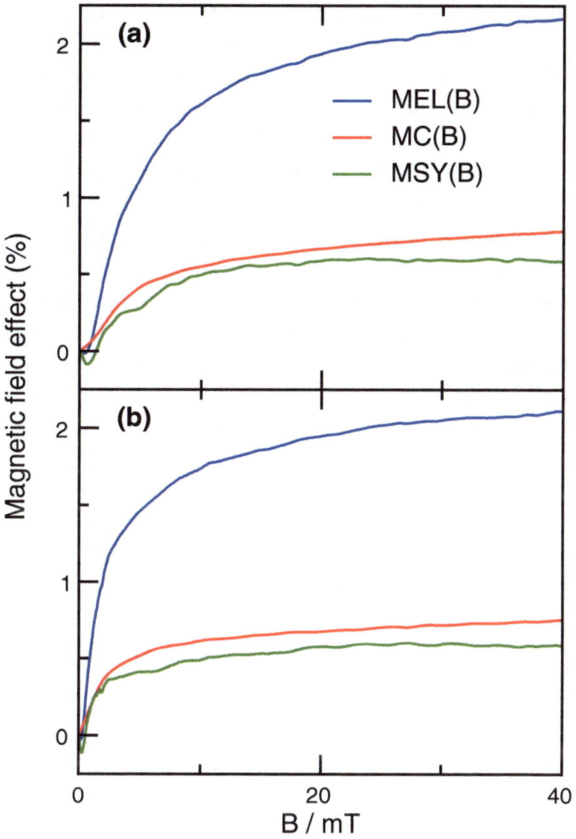

Fig. 6.3 A comparison of the experimental magnetic field effect on the electroluminescence (MEL) and conductance (MC) of **a** H-DOO-PPV and **b** D-DOO-PPV, taken from Refs. [13, 14]. Also shown is the magnetic field effect on the singlet yield (MSY), as defined by the relationship in Eq. (6.18)

simple connection between the experimentally accessible electroluminescence and the singlet exciton yield, Φ_{SE},

$$\mathrm{MEL}(B) \approx \frac{\Phi_{\mathrm{SE}}(B)}{\Phi_{\mathrm{SE}}(0)} - 1. \tag{6.3}$$

Alternatively, since

$$\Phi_{\mathrm{SE}}(B) = \frac{k_{\mathrm{SE}}}{k_{\mathrm{S}}} \Phi_{\mathrm{S}}(B), \tag{6.4}$$

where $\Phi_{\mathrm{S}}(B)$ is the total singlet yield (the fraction of polaron pairs which either recombine to form a singlet exciton or dissociate to give free polarons in the singlet state), Eq. (6.3) may be rewritten as

$$\mathrm{MEL}(B) \approx \mathrm{MSY}(B) = \frac{\Phi_{\mathrm{S}}(B)}{\Phi_{\mathrm{S}}(0)} - 1. \tag{6.5}$$

That is, under the assumption that the number of polaron pairs is not a function of the magnetic field strength, the magnetoelectroluminescence is identical to the magnetic field effect on the overall singlet yield of the polaron pair reaction, $\mathrm{MSY}(B)$.

However, the assumption that n_{PP} is independent of the applied magnetic field strength B is not easily justified; the observation of magnetoconductance in DOO-PPV implies that the number of polarons *is* a function of the applied field. Furthermore, neither quantum mechanical nor semiclassical simulations of the polaron pair model which have relied on this approximation to analyse MEL have been able to even qualitatively reproduce experimental results [13, 16, 17]. $\mathrm{MSY}(B)$ is found to plateau at high fields (see Fig. 6.6 and the discussion in Sect. 1.3.1), while Fig. 6.3 shows that $\mathrm{MEL}(B)$ continues to increase with the magnetic field strength, indicating that the approximate equality in Eq. (6.5) is not reasonable. Therefore, we shall now derive a new expression relating MEL and Φ_{S} without this assumption.

We begin by introducing the normalised density operator of the polaron pair, $\hat{\rho}(B)$, and defining the number density operator as

$$\hat{n}(B) = n_{\mathrm{PP}}(B)\hat{\rho}(B). \tag{6.6}$$

where $n_{\mathrm{PP}}(B)$ is the number of polaron pairs, and

$$\mathrm{tr}\left[\hat{n}(B)\right] = n_{\mathrm{PP}}(B)\mathrm{tr}\left[\hat{\rho}(B)\right] = n_{\mathrm{PP}}(B). \tag{6.7}$$

From Fig. 6.1, the evolution of $\hat{n}(B)$ is given by

$$\frac{\mathrm{d}}{\mathrm{d}t}\hat{n}(B) = -\hat{\hat{L}}(B)\hat{n}(B) + k_{\mathrm{F}}n_{-}(B)n_{+}(B)\hat{\rho}_0, \tag{6.8}$$

with $\hat{\rho}_0 = \hat{1}/4Z$ so that the final term generates triplet and singlet polaron pairs in the correct 3:1 ratio with a first order rate constant k_{F}. $\hat{\hat{L}}(B)$ is the Liouville superoperator defined in Eq. (3.39) which accounts for the spin evolution, recombination, and dissociation processes of the polaron pair, $n_{\pm}(B)$ are the number densities of positively and negatively charged free polarons, $\hat{1}$ is the unit operator on the Hilbert space of the polaron pair, and $4Z = \mathrm{tr}[\hat{1}]$ is the total number of electron and nuclear spin states in this space. Note that $\hat{\hat{L}}(B)$ depends on the magnetic field strength as a consequence of the Zeeman interaction in the Hamiltonian.

At a steady state,

$$\frac{\mathrm{d}}{\mathrm{d}t}\hat{n}(B) = 0, \tag{6.9}$$

and so the total polaron pair number density operator is

$$\hat{n}(B) = k_{\mathrm{F}}\, n_{-}(B)n_{+}(B)\hat{\hat{L}}(B)^{-1}\hat{\rho}_0. \tag{6.10}$$

The number of singlet polaron pairs, $n_{\mathrm{S}}(B)$, is therefore simply

$$\begin{aligned} n_S(B) &= \mathrm{tr}\left[\hat{P}_S \hat{n}(B)\right] \\ &= \frac{k_F}{k_S} n_-(B) n_+(B)\, \Phi_S(B), \end{aligned} \tag{6.11}$$

where $\hat{P}_S$ is the singlet projection operator defined in Eq. (2.32), $k_S = k_{SE} + k_D$ is the first order rate constant for loss of population of singlet polaron pair, and

$$\Phi_S(B) = k_S\, \mathrm{tr}[\hat{P}_S \hat{\hat{L}}(B)^{-1} \hat{\rho}_0], \tag{6.12}$$

which we demonstrated in Sect. 3.2 is precisely the singlet yield of the polaron pair recombination reaction.

From Fig. 6.1, the steady state electroluminescence must clearly be proportional to $k_{SE}\, n_S(B)$, and so the magnetic field effect on the luminescence is given by

$$\mathrm{MEL}(B) = \frac{n_-(B) n_+(B)\, \Phi_S(B)}{n_-(0) n_+(0)\, \Phi_S(0)} - 1. \tag{6.13}$$

If we now assume that electrons are injected into and removed from the polymer film through Ohmic contacts, then there will only be a very small net space charge within the film [34],

$$\begin{aligned} \delta n(B) &= n_+(B) - n_-(B) \approx 0, \\ n_\pm(B) \approx n(B) &= \frac{1}{2}\left(n_+(B) + n_-(B)\right). \end{aligned} \tag{6.14}$$

This has been shown to be a reasonable assumption in oLEDs based on PPV derivatives which use calcium cathodes and poly(3,4-ethylenedioxythiophene)-poly(styrenesulfonate) (PEDOT-PSS) hole injecting layers [11], like those employed in the experiments of Nguyen et al. [13, 14] With this assumption,

$$\mathrm{MEL}(B) = \frac{n(B)^2\, \Phi_S(B)}{n(0)^2\, \Phi_S(0)} - 1. \tag{6.15}$$

Using this assumption and the further assumption that the mobilities of the positive and negative free polarons, μ_+ and μ_-, are independent of the applied magnetic field strength, we may write an expression for MC:

$$\begin{aligned} \mathrm{MC}(B) &= \frac{\mu_+ n_+(B) + \mu_- n_-(B)}{\mu_+ n_+(0) + \mu_- n_-(0)} - 1 \\ &\approx \frac{n(B)}{n(0)} - 1. \end{aligned} \tag{6.16}$$

Combining Eqs. (6.15) and (6.16) we find an expression for MEL in terms of MSY and MC:

$$\mathrm{MEL}(B) + 1 = (\mathrm{MC}(B) + 1)^2\,(\mathrm{MSY}(B) + 1)\,, \tag{6.17}$$

where MSY has been defined in Eq. (6.5).

This relationship between MEL(B), MC(B), and MSY(B) has been derived making only two assumptions: firstly, that the contacts are approximately Ohmic, and secondly that the mobilities μ_+ and μ_- of P^+ and P^- are independent of the applied magnetic field. Given these assumptions, the result is a direct consequence of the polaron pair mechanism for magnetoelectroluminescence illustrated in Fig. 6.1. Notice in particular that we have not made any assumptions about the mechanism of the magnetoconductance, other than that the applied magnetic field does not change the mobilities of the free polarons. This approximation is consistent with many of the mechanisms that have been proposed for magnetoconductance in polymer films, including the polaron pair mechanism [27], the bipolaron mechanism in the regime of positive MC [31, 35], and a more recently suggested mechanism involving the trapping of polarons at defect sites [33]. In all three of these scenarios, the magnetic field effect on the conductance is believed to arise from a change in the free polaron number density in the presence of the magnetic field rather than a change in the free polaron mobility. Indeed, it is explicitly stated in Ref. [27] that no magnetic field effect is found experimentally on the mobilities of the free polarons in PPV derivatives.

The curves labelled MSY(B) in Fig. 6.3 were obtained from the experimental MEL(B) and MC(B) curves by rearranging Eq. (6.17) into the form

$$\mathrm{MSY}(B) = \frac{\mathrm{MEL}(B) + 1}{(\mathrm{MC}(B) + 1)^2} - 1. \tag{6.18}$$

Therefore, another way of viewing Eq. (6.17) is that it provides a link between the experimentally measurable magnetoelectroluminescence and magnetoconductance and the theoretically calculable magnetic field effect on the singlet yield of the polaron pair recombination reaction. We shall now take this view, and in the next section calculate $\Phi_S(B)$, and hence MSY(B), for DOO-PPV.

6.3 Simulating the Singlet Yield

We have shown in Sect. 3.2 how the singlet yield defined in Eq. (6.12) may be straightforwardly evaluated within Schulten–Wolynes theory,

$$\Phi_S(B) = \int d\mathbf{I}_1\, g(\mathbf{I}_1) \int d\mathbf{I}_2\, g(\mathbf{I}_2)\, k_S\, \mathrm{tr}_S[\hat{P}_S \hat{\hat{L}}(B, \mathbf{I}_1, \mathbf{I}_2)^{-1} \hat{\rho}_0]. \tag{6.19}$$

Since each polaron contains $N_i \sim 100$ nuclear spins, we expect SW theory to be reasonably accurate for this problem. In order to use this expression, we need to define the distribution of the magnitude of the resultant vector of the nuclear spins,

$g(\mathbf{I}_i)$. It is convenient here to express $g(\mathbf{I}_i)$ in terms of the effective hyperfine field,

$$B_{\mathrm{hyp},i} = \sqrt{\sum_{k=1}^{N_i} a_{ik}^2 I_{ik}(I_{ik}+1)}, \tag{6.20}$$

so that the distribution becomes

$$g(\mathbf{I}_i) = \left(\frac{3}{2\pi B_{\mathrm{hyp},i}}\right)^{3/2} e^{-3I_i^2/2B_{\mathrm{hyp},i}^2}. \tag{6.21}$$

Therefore, within SW theory the hyperfine couplings of a polaron in DOO-PPV reduce to a single parameter, $B_{\mathrm{hyp},i}$.

6.3.1 Hyperfine Fields in DOO-PPV

The hyperfine field of a polaron in DOO-PPV may be found from ESR experiments. In fact, this is all that may be deduced from such experiments, since the spectra do not exhibit resolved hyperfine splittings [36, 37]. Instead, the integrated ESR signal has a broad approximately Gaussian lineshape, whose full width at half maximum (FWHM) is related to $B_{\mathrm{hyp},i}$ by

$$\mathrm{FWHM} = \sqrt{\frac{8\ln 2}{3}} B_{\mathrm{hyp},i}, \tag{6.22}$$

as expected from Eq. (6.21). Kuroda et al. [36] have measured the light-induced electron spin resonance spectra of thin films of two different dialkoxy derivatives of PPV (MEH-PPV and CN-PPV). They found a FWHM of 0.66 mT for one and 0.45 mT for the other, with no evidence for any difference between the contributions of positive and negative polarons to either ESR signal. More recently, Zezin et al.[37] have measured an ESR linewidth of 0.5 mT for the positive polarons in long oligomers of DOO-PPV in an irradiated glassy toluene solution at 77 K in the presence of an electron scavenger, and performed similar experiments in the presence and absence of the scavenger to deduce that the positive and negative polarons of MEH-PPV have somewhat different ESR linewidths (0.37 and 0.59 mT, respectively).

Since all of these linewidths are fairly similar, and since the results we shall present below are fairly insensitive to the precise choice of $B_{\mathrm{hyp},i}$, we shall avoid introducing too many free parameters into our calculations by assuming a FWHM ESR linewidth of 0.5 mT for both the positive and the negative polarons in thin films of H-DOO-PPV. The use of 0.5 mT for the positive polarons is consistent with the long DOO-PPV oligomer experiments of Zezin et al. [37] The assumption that this is the same for both positive and negative polarons is consistent with the thin film experiments of Kuroda et al. [36] and with the particle-hole symmetry of the Pariser-Parr-Pople

Hamiltonian for PPV that was used to interpret their results [38]. Strictly speaking, particle-hole symmetry will be broken by the electronegative oxygen atoms that are α to the conjugated framework of DOO-PPV. However, explicit density functional theory calculations on short oligomers of a more manageable compound suggest that this is only a minor effect: the computed hyperfine field strengths $B_{\text{hyp},i}$ of the positively and negatively charged n = 3 oligomers of DOE-PPV (where E = ethyl) differ by less than 10%.

According to Eq. (6.22), a FWHM of 0.5 mT gives $B_{\text{hyp}}(H) = 0.37\,\text{mT}$ for H-DOO-PPV, and the corresponding parameter for D-DOO-PPV can be worked out as follows. Neglecting ^{13}C nuclei, which will be present throughout the polymer with just ~1% natural abundance [39], the only magnetic nuclei in H-DOO-PPV are protons. Since H has $I = 1/2$ and D has $I = 1$, and the hyperfine coupling constants a_{ik} in Eq. (6.20) are proportional to the gyromagnetic ratios γ_H and γ_D of the two nuclei, the effect of deuteration will be to reduce B_{hyp} to

$$B_{\text{hyp}}(D) = \frac{\gamma_D}{\gamma_H}\sqrt{\frac{8}{3}}B_{\text{hyp}}(H) \approx \frac{1}{4}B_{\text{hyp}}(H), \tag{6.23}$$

which gives $B_{\text{hyp}}(D) = 0.093\,\text{mT}$. This assumes that D-DOO-PPV is fully deuterated. In fact, the material used in Refs. [13, 14] was only partially deuterated: the H atoms directly bonded to the conjugated π system were deuterated whereas those in the DOO side chains were not. However, the explicit DFT calculations on $n = 3$ oligomers of DOE-PPV indicate that the hyperfine couplings in the side chains make up less than 0.25% of B_{hyp}, as one would expect on the basis of chemical intuition. Therefore, we will neglect their contribution and take the ratio B_{hyp}(H)/B_{hyp}(D) to be 4.

While direct ESR measurements of the hyperfine fields in thin films of DOO-PPV have not been performed, optically detected magnetic resonance (ODMR) experiments have [14]. However, these experiments produce complicated lineshapes which are open to multiple interpretations. For example, some studies assume that the polarons have significantly different hyperfine fields and fit the lineshape to two Gaussians [12, 35, 40]. On the other hand, the authors who measured the ODMR in DOO-PPV fit the data to a Voigt profile, which is the convolution of a Gaussian lineshape to account for the distribution of the hyperfine interactions and a Lorentzian lineshape to account for lifetime broadening [14]. Therefore, we prefer to use ESR experiments performed on thin films of similar dialkoxy PPVs and on solutions of oligomers of DOO-PPV to estimate the hyperfine fields of the polarons in DOO-PPV, which are straightforward to interpret having almost perfectly Gaussian lineshapes. Nevertheless, we can show that our ESR-based values of B_{hyp} are consistent with the ODMR experiments. Figure 6.4 shows that the ODMR signals of both H-DOO-PPV and D-DOO-PPV from Ref. [14] can be fit to Voigt profiles [41],

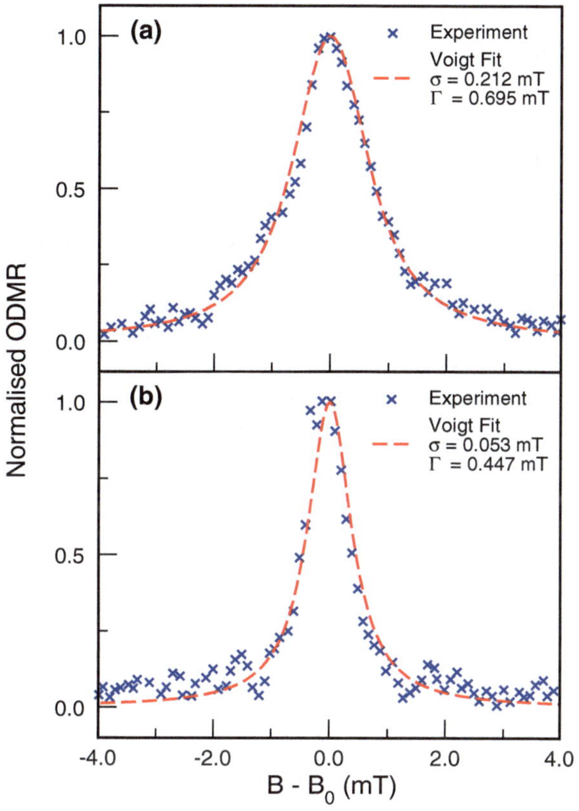

Fig. 6.4 Voigt fits to the ODMR signals of **a** H-DOO-PPV and **b** D-DOO-PPV. The Voigt profile is defined in Eq. (6.24), with σ determined by the B_{hyp} values taken from ESR experiments. The experimental data is taken from Ref. [14]

$$V(B;\sigma,\Gamma) = \int_{\infty}^{\infty} G(B';\sigma)\, L(B-B';\Gamma)\mathrm{d}B',$$

$$G(x;\sigma) = \frac{e^{-x^2/2\sigma^2}}{\sigma\sqrt{2\pi}},$$

$$L(x;\Gamma) = \frac{\Gamma}{\pi\left(x^2+\Gamma^2\right)}, \tag{6.24}$$

where the standard deviation of the Gaussian distribution, σ, is chosen to correspond to the hyperfine fields of $B_{\text{hyp}}(H) = 0.37\,\text{mT}$ and $B_{\text{hyp}}(D) = 0.093\,\text{mT}$, and the scale parameters of the Lorentzian components of the Voigt profiles, Γ, are taken to be free parameters.

While Fig. 6.4 demonstrates that the hyperfine fields deduced from the ESR experiments are consistent with the ODMR data, there are other combinations of σ and Γ which produce a fit of similar quality. Since the lifetime broadening is difficult to quantify, this makes it very difficult to extract a unique value of σ and therefore B_{hyp} from ODMR experiments alone. This is the main reason for our preference for deducing the hyperfine interactions from ESR experiments on stable polarons, in which lifetime broadening is entirely avoided.

6.3.2 Singlet-Triplet Dephasing

Careful inspection of Fig. 6.3 reveals that the hyperfine interactions alone are insufficient to explain the MSY obtained using Eq. (6.18) and the experimentally measured MEL and MC of H-DOO-PPV and D-DOO-PPV. Both curves tend towards the same asymptote, but they have very different values of $B_{1/2}$, the magnetic field strength at which $\mathrm{MSY}(B)$ is half of $\mathrm{MSY}(\infty)$ introduced in Sect. 1.3.1. If the spin evolution of the polaron pair were governed solely by the Zeeman and hyperfine interactions, we would expect $B_{1/2}$ to conform to the Weller equation introduced in Sect. 1.3.1 [42],

$$B_{1/2} = 2\frac{B_{\mathrm{hyp},1}^2 + B_{\mathrm{hyp},2}^2}{B_{\mathrm{hyp},1} + B_{\mathrm{hyp},2}}. \tag{6.25}$$

In the case that the positive and negative polarons have the same hyperfine fields, this reduces to

$$B_{1/2} = 2B_{\mathrm{hyp}}. \tag{6.26}$$

However, plotting the magnetic field effect on the singlet yield on an expanded scale in Fig. 6.5 clearly shows that the values of $B_{1/2}$ in H-DOO-PPV and D-DOO-PPV are not consistent with the Weller equation. Both $B_{1/2}(H) = 5.3\,\mathrm{mT}$ and $B_{1/2}(D) = 2.0\,\mathrm{mT}$ are an order of magnitude larger than our estimates of the hyperfine fields strengths, $B_{\mathrm{hyp}}(H) = 0.37\,\mathrm{mT}$ and $B_{\mathrm{hyp}}(D) = 0.093\,\mathrm{mT}$. Furthermore, the ratio $B_{1/2}(H)/B_{1/2}(D) = 2.65$ is significantly smaller than the factor of 4 predicted by Eq. (6.23).

Clearly some additional phenomenon plays an important role in the spin dynamics of the polaron pairs in these oLEDs. A number of processes have already been considered and discounted [16], but relatively little attention has been given to the role of electron spin relaxation. This is perhaps unsurprising, since the most common mechanism of relaxation, the modulation of the hyperfine interactions, leads to extremely long relaxation times of polarons in solid state organic polymers [43, 44]. However, other relaxation mechanisms may be significant, such as modulation of the exchange interaction between the two electrons in the polaron pair. As discussed in Sect. 2.1.1, the strength of the exchange interaction depends exponentially on the separation between the two spins, which varies due to the migration of the polarons within the pair along their polymer chain [45, 46]. This modulation causes singlet-triplet dephasing, introduced in Sect. 2.7.2, which in other contexts has been shown to lead to $B_{1/2}$ values significantly larger than those predicted by the Weller equation [47].

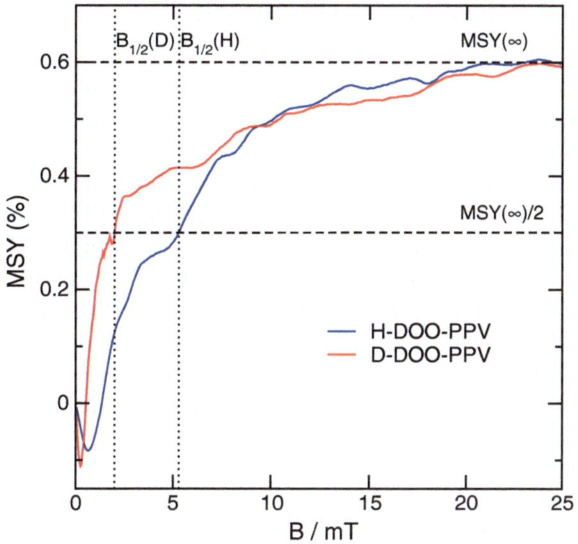

Fig. 6.5 The magnetic field effect on the singlet yield of the polaron pair recombination reaction in both H-DOO-PPV and D-DOO-PPV, with the values of MSY(∞), $B_{1/2}(H)$, and $B_{1/2}(D)$ highlighted

This singlet triplet dephasing may be easily accounted for in our semiclassical calculations, since it is phenomenologically described by Eq. (2.74). Adding this term to the Liouvillian defined in Eq. (3.39) gives

$$-\hat{\hat{L}}(B, \mathbf{I}_1, \mathbf{I}_2)\hat{\rho} = -i[\hat{H}(B, \mathbf{I}_1, \mathbf{I}_2), \hat{\rho}] - \{\hat{K}, \hat{\rho}\} - k_R\left(\hat{P}_S\hat{\rho}\hat{P}_T + \hat{P}_T\hat{\rho}\hat{P}_S\right), \tag{6.27}$$

where k_R is a singlet-triplet dephasing rate constant, the recombination operator $\hat{K}$ is described in Sect. 2.2.1, and within SW theory the Hamiltonian is

$$\hat{H}(B, \mathbf{I}_1, \mathbf{I}_2) = (\boldsymbol{\omega}_1 + \mathbf{I}_1) \cdot \hat{S}_1 + (\boldsymbol{\omega}_2 + \mathbf{I}_2) \cdot \hat{S}_2 \tag{6.28}$$

It should be noted that the average exchange interaction has been neglected in this Hamiltonian, even though its modulation is included in the final term of Eq. (6.27). There are three reasons for this. Firstly, the average strength of the exchange coupling constant is extremely difficult to determine in these organic polymers [48], and we do not want to add any additional parameters to our simulation unless absolutely necessary. Secondly, the exponential dependence of the exchange coupling on polaron separations suggests that even if the average exchange coupling is very small, brief periods of time during which the polarons are in close proximity would cause a large modulation effect. Finally, we found that the spin dynamics of the problem were largely unaffected by the inclusion of a physically reasonably constant exchange coupling. Therefore, we have neglected the average coupling for simplicity.

6.4 Results and Discussion

We are now in a position to simulate the experimental results in Fig. 6.3. This requires us to find three free parameters: the singlet and triplet decay rate constants, k_S and k_T, and the dephasing rate constant, k_R, which we shall assume are all unaffected by deuteration. We therefore performed a least squares fit to the experimental MSY(B) of both deuterated and undeuterated DOO-PPV simultaneously in the $\{k_S, k_T, k_R\}$ parameter space using the downhill simplex method. This found a single minimum at $k_S = 4.84 \times 10^5\ \text{s}^{-1}$, $k_T = 4.94 \times 10^5\ \text{s}^{-1}$, and $k_R = 1.50 \times 10^8\ \text{s}^{-1}$. The resulting simulated MSY(B) curves are compared to the experimental magnetic field effect in Fig. 6.6. The agreement between theory and experiment is clearly excellent for H-DOO-PPV at all but the very lowest magnetic field strengths. The agreement is also good for D-DOO-PPV, although the experimental curve does have a kink between $B = 2.5$ and 20 mT that is not captured by the theoretical calculation. Assuming that this can be dismissed as an artefact, the agreement between theory and experiment is clearly very good for both isotopologues of DOO-PPV over a wide range of applied magnetic fields.

Figure 6.6 both justifies our expression for MEL(B) in Eq. (6.17) and shows that singlet-triplet dephasing plays an important role in the polaron pair spin dynamics. If, as has been done in the past [13, 16, 17], Eq. (6.5) is used to fit the computed MSY(B) to the experimental MEL(B) by optimising the same three parameters, k_R, k_S, and k_T, the results show nowhere near as good agreement as the fits in Fig. 6.6. In addition, if k_R is set to zero and singlet-triplet dephasing is not included, it is not possible to reproduce the $B_{1/2}$ values of the experimental MEL(B) curves without invoking infeasibly large and unrelated hyperfine fields $B_{\text{hyp}}(H)$ and $B_{\text{hyp}}(D)$.

Our simulations do not capture the experimentally observed low field effect (LFE) [49] – more commonly referred to as an ultra-small magnetic field effect (USMFE) in the context of organic semiconductor devices [16, 50] – which results in a negative dip in MSY(B) below 1.3 mT for H-DOO-PPV and 0.5 mT for D-DOO-PPV. This LFE has been captured in earlier calculations [13, 16, 17], and we too have found that it can be captured by optimising the parameters k_R, k_S, and k_T to fit just the low field region. However, the resulting MSY(B) curves do not give nearly such good agreement with experiment over the full range of magnetic field strengths as those in Fig. 6.6. It should also be noted that the Schulten–Wolynes approximation is expected to be least reliable in the low field region, as discussed in Sect. 3.3.

We found that in order to fit our simulations to the experimental data k_S was required to be smaller than k_T. This is at odds with theoretical calculations of the recombination rate of polaron pairs [29], but is demanded by the experimental data, since MSY(B) is positive at high fields [17]. When $k_T = k_S$, $\Phi_S(B) = 1/4$ for all B, and therefore MSY$(\infty) = 0$. However, when $k_T > k_S$, the triplet states of the polaron pair decay more rapidly than the singlet state, and $\Phi_S(B)$ is less than $1/4$. The decrease in $\Phi_S(B)$ is most pronounced at low field strengths, where singlet population can be lost through intersystem crossing to all three components of the triplet, as described in Sect. 1.3.2. At high field strengths, the Zeeman splittings of the $T_\pm$ triplet components

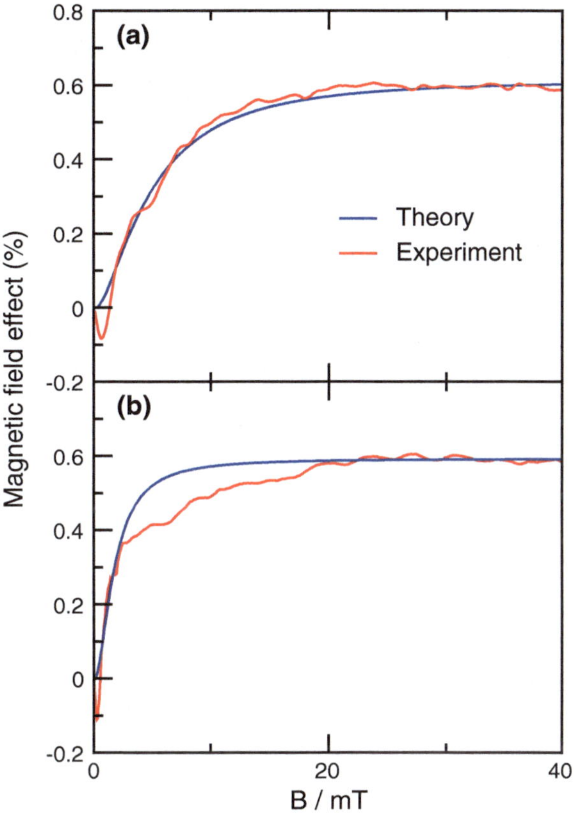

Fig. 6.6 Comparison of the simulated and experimental magnetic field effect on the singlet yield (MSY) of **a** H-DOO-PPV and **b** D-DOO-PPV. The experimental curve was obtained from Eq. (6.18) using the magnetoelectroluminescence and magnetoconductance data reported in Refs. [13, 14]

make them energetically inaccessible to singlet-triplet interconversion, so singlet population can only be lost through interconversion to T_0, as discussed in Sect. 1.3.1. Therefore $\Phi_S(0) < \Phi_S(\infty) < 1/4$ and $\text{MSY}(\infty) > 0$. Conversely, when $k_T < k_S$, a similar argument gives $\Phi_S(0) > \Phi_S(\infty) > 1/4$ and $\text{MSY}(\infty) < 0$.

Finally, while we have deliberately avoided making any assumptions about the mechanism of the magnetoconductance in DOO-PPV, our results do in fact shed some light on this. One of the most widely discussed mechanisms of magnetoconductance is the polaron pair mechanism, in which the effect of the magnetic field on the conductance is assumed to arise from its effect on the dissociation yield of the polaron pair back to free charge carriers (P^+ and P^-) [16, 27]. However, it is clear from Fig. 6.3 that the magnetic field effect on the overall singlet yield of the polaron pair recombination reaction, which includes both the singlet exciton yield and the yield of free charge carriers in the singlet state, has already saturated at a field strength of 20 mT, whereas the magnetoelectroluminescence and magnetoconductance continue to increase at higher field strengths (up to 40 mT). Since the overall triplet yield (including the yield of free charge carriers in the triplet state) must also have saturated when the overall singlet yield has saturated, this clearly implies that the change in the magnetoconductance beyond 20 mT cannot come from the polaron pair mechanism. So although it seems likely, given the similarities between the MC(B) and MSY(B) curves in Fig. 6.3, that the polaron pair mechanism of magnetoconductance does play an important role in DOO-PPV, there must also be some other mechanism in operation at high magnetic field strengths.

6.5 Conclusion

In this chapter, we have used the polaron pair mechanism outlined in Fig. 6.1 to derive an expression in Eq. (6.17) which relates the theoretically calculable magnetic field effect on the singlet yield of the polaron pair recombination reaction, MSY(B), to the experimentally accessible magnetoelectroluminescence MEL(B) and magnetoconductance MC(B) of the polymer film. We have argued that this expression holds independently of the mechanism of the magnetoconductance, provided that the mobilities of the positive and negative polarons are not affected by the magnetic field.

We have also discussed the parameters required for a semiclassical calculation of MSY(B) in polymers such as DOO-PPV, the standard deviations of the hyperfine fields in which can be extracted from ESR linewidth measurements. Once these hyperfine fields are known, the calculation of MSY(B) involves just three empirical parameters: the overall singlet and triplet decay rate constants k_S and k_T of the polaron pair and a singlet-triplet dephasing rate constant k_R. The last of these parameters has not been included in any previous theory of magnetoelectroluminescence, but we have argued on the basis of the Weller equation [42] that its inclusion is necessary to reproduce the experimental results of Nguyen et al. for H-DOO-PPV and D-DOO-PPV [13, 14]. The fits to the experimental data obtained from these calculations are shown in Fig. 6.6. The agreement between theory and experiment is very good for both H-DOO-PPV and D-DOO-PPV over a wide range of magnetic field strengths.

References

1. Burroughes, J. H. et al. (1990). Light-emitting diodes based on conjugated polymers. *Nature, 347*, 539–541.
2. Yang, Y. (1997). Polymer electroluminescent devices. *Materials Research Society Bulletin, 22*, 31–38.
3. Hoofman, R. J. O. M., de Haas, M. P., Siebbeles, L. D. A., & Warman, J. M. (1998). Highly mobile electrons and holes on isolated chains of the semiconducting polymer poly (phenylene vinylene). *Nature, 392*, 54–56.
4. Grozema, F. C. et al. (2002). Theoretical and experimental studies of the opto-electronic properties of positively charged oligo (phenylene vinylene)s: Effects of chain length and alkoxy substitution. *Journal of Chemical Physics, 117*, 11366.
5. Tang, C. W., & Vanslyke, S. A. (1987). Organic electroluminescent diodes. *Applied Physics Letters, 51*, 913–915.

6. Kido, J. (1999). Organic displays. *Physics World*, *12*, 27–30.
7. Forrest, S. R. (2004). The path to ubiquitous and low-cost organic electronic appliances on plastic. *Nature*, *428*, 911–918.
8. (2009). OLED displays and organic photovoltaics. *Nature Photonics*, *3*, 457.
9. Lupton, J. M., McCamey, D. R., & Boehme, C. (2010). Coherent spin manipulation in molecular semiconductors: Getting a handle on organic spintronics. *ChemPhysChem*, *11*, 3040–3058.
10. Reineke, S. (2015). Complementary LED technologies. *Nature Materials*, *14*, 459–462.
11. Leger, J. M., Carter, S. A., Ruhstaller, B., Scherf, U., & Tillman, H. (2003). Thickness-dependent changes in the optical properties of PPV- and PF-based polymer light emitting diodes. *Physical Review B*, *68*, 054209.
12. McCamey, D. R. et al. (2010). Hyperfine-field-mediated spin beating in electrostatically bound charge carrier pairs. *Physical Review Letters*, *104*, 017601.
13. Nguyen, T. D., Gautam, B. R., Ehrenfreund, E., & Vardeny, Z. V. (2010). Magnetoconductance response in unipolar and bipolar organic diodes at ultrasmall fields. *Physical Review Letters*, *105*, 166804.
14. Nguyen, T. D. et al. (2010). Isotope effect in spin response of pi-conjugated polymer films and devices. *Nature Materials*, *9*, 345–352.
15. Janssen, P. et al. (2011). On the role of minority carriers in the frequency dependence of organic magnetoresistance. *Synthetic Metals*, *161*, 617–621.
16. Ehrenfreund, E., & Vardeny, Z. V. (2012). Effects of magnetic field on conductance and electroluminescence in organic devices. *Israel Journal of Chemistry*, *52*, 552–562.
17. Kersten, S. P., Schellekens, A. J., Koopmans, B., & Bobbert, P. A. (2011). Magnetic-field dependence of the electroluminescence of organic light-emitting diodes: A competition between exciton formation and spin mixing. *Physical Review Letters*, *106*, 197402.
18. McCamey, D. R., Lee, S. Y., Paik, S. Y., Lupton, J. M., & Boehme, C. (2010). Spin-dependent dynamics of polaron pairs in organic semiconductors. *Physical Review B*, *82*, 125206.
19. Wang, F., Yang, C. G., Ehrenfreund, E., & Vardeny, Z. V. (2010). Spin dependent reactions of polaron pairs in PPV-based organic diodes. *Synthetic Metals*, *160*, 297–302.
20. Landau, L. D. (1933). Über die Bewegung der Elektronen im Kristallgitter. *Physikalische Zeitschrift der Sowjetunion*, *3*, 664.
21. Tozer, O. R., & Barford, W. (2014). Localization of large polarons in the disordered Holstein model. *Physical Review*, *89*, 155434.
22. Marcus, M., Tozer, O. R., & Barford, W. (2014). Theory of optical transitions in conjugated polymers. II. Real systems. *Journal of Chemical Physics*, *141*, 164102.
23. Anderson, P. W. (1958). Absence of diffusion in certain random lattices. *Physical Review*, *109*, 1492–1505.
24. Mott, N. F., & Twose, W. (1961). The theory of impurity conduction. *Advances in Physics*, *10*, 107.
25. Hsu, J. W. P., Yan, M., Jedju, T. M., Rothberg, L. J., & Hsieh, B. R. (1994). Assignment of the picosecond photoinduced absorption in phenylene vinylene polymers. *Physical Review B*, *49*, 712–715.
26. Mizes, H. A., & Conwell, E. M. (1994). Photoinduced charge transfer in poly (p-phenylene vinylene). *Physical Review B*, *50*, 243–246.
27. Frankevich, E. L. et al. (1992). Polaron-pair generation in poly (phenylene vinylenes). *Physical Review B*, *46*, 9320–9324.
28. Dyakonov, V., Röosler, G., Schwoerer, M., & Frankevich, E. L. (1997). Evidence for triplet interchain polaron pairs and their transformations in polyphenylenevinylene. *Physical Review B*, *56*, 3852–3862.
29. Barford, W. (2004). Theory of singlet exciton yield in light-emitting polymers. *Physical Review B*, *70*, 205204.
30. Hu, B., & Wu, Y. (2007). Tuning magnetoresistance between positive and negative values in organic semiconductors. *Nature Materials*, *6*, 985–91.
31. Bobbert, P. A., Nguyen, T. D., Van Oost, F. W. A., Koopmans, B., & Wohlgenannt, M. (2007). Bipolaron mechanism for organic magnetoresistance. *Physical Review Letters*, *99*, 216801.

32. Lupton, J. M., & Boehme, C. (2008). Magnetoresistance in organic semiconductors. *Nature Materials, 7*, 598.
33. Cox, M. et al. (2014). Spectroscopic evidence for trap-dominated magnetic field effects in organic semiconductors. *Physical Review B, 90*, 155205.
34. Parmenter, R. H., & Ruppel, W. (1959). Two-carrier space-charge-limited current in a trap-free insulator. *Journal of Applied Physics, 30*, 1548–1558.
35. Behrends, J. et al. (2010). Bipolaron formation in organic solar cells observed by pulsed electrically detected magnetic resonance. *Physical Review Letters, 105*, 176601.
36. Kuroda, S. et al. (2000). Spin distributions and excitation spectra of optically generated polarons in poly (p-phenylenevinylene) derivatives. *Chemical Physics Letters, 325*, 183–188.
37. Zezin, A. A., Feldman, V. I., Warman, J. M., Wildeman, J., & Hadziioannou, G. (2004). EPR study of positive holes on phenylene vinylene chains: From dimer to polymer. *Chemical Physics Letters, 389*, 108–112.
38. Shimoi, Y., Abe, S., Kuroda, S.-I., & Murata, K. (1995). Polarons and their ENDOR spectra in poly (p-phenylene vinylene). *Solid State Communications, 95*, 137–141.
39. Rosman, K. J. R., & Taylor, P. D. P. (1998). Isotopic compositions of the elements 1997. *Pure and Applied Chemistry, 70*, 217–235.
40. Baker, W. J., Keevers, T. L., Lupton, J. M., McCamey, D. R., & Boehme, C. (2012). Slow hopping and spin dephasing of coulombically bound polaron pairs in an organic semiconductor at room temperature. *Physical Review Letters, 108*, 267601.
41. Langford, J. I. (1978). A rapid method for analysing the breadths of diffraction and spectral lines using the voigt function. *Journal of Applied Crystallography, 11*, 10–14.
42. Weller, A., Nolting, F., & Staerk, H. (1983). A quantitative interpretation of the magnetic field effect on hyperfine-coupling-induced triplet fromation from radical ion pairs. *Chemical Physics Letters, 96*, 24–27.
43. Reufer, M. et al. (2005). Spin-conserving carrier recombination in conjugated polymers. *Nature Materials, 4*, 340–346.
44. Yang, C. G., Ehrenfreund, E., & Vardeny, Z. V. (2007). Polaron spin-lattice relaxation time in pi-conjugated polymers from optically detected magnetic resonance. *Physical Review Letters, 99*, 157401.
45. Deotare, P. B. et al. (2015). Nanoscale transport of charge-transfer states in organic donor-acceptor blends. *Nature Materials, 14*, 1130–1134.
46. Lee, C. K., Shi, L., & Willard, A. P. (2016). A model of charge-transfer excitons: Diffusion, spin dynamics, and magnetic field effects. *Journal of Physical Chemistry Letters, 7*, 2246–2251.
47. Maeda, K. et al. (2012). Magnetically sensitive light-induced reactions in cryptochrome are consistent with its proposed role as a magnetoreceptor. *Proceedings of the National Academy of Sciences of the United States of America, 109*, 4774–4779.
48. Lane, P. A., Wei, X., & Vardeny, Z. V. (1997). Spin and spectral signatures of polaron pairs in π-conjugated polymers. *Physical Review B, 56*, 4626–4637.
49. Timmel, C. R., Till, U., Brocklehurst, B., Mclauchlan, K. A., & Hore, P. J. (1998). Effects of weak magnetic fields on free radical recombination reactions. *Molecular Physics, 95*, 71–89.
50. Nguyen, T. D., Ehrenfreund, E., & Vardeny, Z. V. (2013). Organic magneto-resistance at small magnetic fields; compass effect. *Organic Electronics: Physics, Materials, Applications, 14*, 1852–1855.

Chapter 7
Conclusions and Further Work

There has been significant interest in radical pair reactions for a number of years [1–5], due to their relevance to a number of biological and technological systems [6–11]. In particular, the effect of magnetic fields on radical pair reactions has been widely studied [12–16], with even very weak magnetic interactions able to dramatically change the rate and yield of a reaction [17, 18]. In this thesis, we have developed both quantum mechanical and semiclassical methods of simulating radical pair reactions, and then applied those methods to three real systems in order to obtain some physical insight into their behaviour. Here we shall summarise our findings, before suggesting two areas for further work where the application of the semiclassical theory introduced in Sect. 3.1 has shown promising early results.[1]

7.1 Conclusions

7.1.1 Theory

The quantum mechanical description of a radical pair reaction is relatively straightforward, and is outlined in Chap. 2. However, fully quantum mechanical simulations of realistic radical pairs are very difficult to perform, since the Hilbert space of the radical pair grows exponentially with the number of nuclear spins in the radical pair. The CPU time required for a deterministic evaluation of the ensemble average of an observable scales as $Z^2 \log Z$, where Z is the total number of nuclear spin states of the radical pair. Therefore, the goal of Chap. 2 was to find a quantum mechanical method which improves upon this scaling, and in Chap. 3 we set out to develop approximate

[1]Parts of this chapter have been reproduced with permission from Lewis, A. M. et al. *Journal of Chemical Physics* **149**, 034103 (2018), https://aip.scitation.org/doi/10.1063/1.5038558.

A. Lewis, *Spin Dynamics in Radical Pairs*, Springer Theses,
https://doi.org/10.1007/978-3-030-00686-0_7

semiclassical methods which avoided the exponential scaling altogether. These new methods allow us to simulate the spin dynamics of realistic systems.

In Sect. 2.5 we demonstrated that the computational time of a fully quantum mechanical simulation can be reduced by a factor of $\mathcal{O}(Z)$ by using a stochastic method. We introduced coherent spin states, and showed that the ensemble average of an observable of the radical pair may be written as a multidimensional integral over the coherent spin state basis of each nuclear spin. These integrals may be evaluated by Monte Carlo sampling of the directions of the initial coherent nuclear spin states. If the number of samples required to converge the integrals, M, is smaller than Z, then this method will be significantly faster than a deterministic evaluation. We demonstrated that for a model radical pair which included 20 $I = 1/2$ nuclear spins, only $M = 200$ samples are required to obtain converged results over a wide range of applied field strengths. Since the total number of nuclear spin states $Z = 1,048,576$, the stochastic calculation is over 5000 times faster than deterministic one would be in this case.

The reason for the efficiency of this method lies in the fact that the observables of interest in radical pairs are typically probabilities or yields. Since these are bounded between zero and one, the standard deviation can be at most 1/2, and is often much smaller. As a result, only a small number of samples are required to obtain well converged results. In addition, when evaluating an ensemble average stochastically it is advantageous to sample coherent spin states, rather than eigenstates of the total spin projection operator, $\hat{J}_z$. When anisotropic interactions are neglected, the spin Hamiltonian commutes with $\hat{J}_z$, which divides the Hilbert space into independent sectors of different M_J, with the evolution of any eigenstate of $\hat{J}_z$ confined to a single one of these sectors. However, each coherent spin state contains a contribution from each of these sectors, so sampling these states leads to a faster convergence of ensemble average.

In Sect. 3.1 we introduced our semiclassical theory, which eliminates the exponential scaling problem altogether. This method treats the electron and nuclear spins as classical vectors, and derives equations of motion for these vectors from the Heisenberg equations of motion for the corresponding quantum mechanical operators. We found that a small modification to the equations of motion of the nuclear spins is required in order to produce the correct behaviour in the limit of symmetric recombination rates, but that this correction is both unique and physically motivated. This produces a set of $3N + 16$ coupled differential equations, which can be solved numerically. As a result, the computational time required for semiclassical simulations scales only linearly with the number of nuclear spins in the radical pair, rather than exponentially. We found that for radical pairs which contain $N > 12$ nuclear spins, the results of the semiclassical theory are in good agreement with exact quantum mechanical calculations, and are expected to become increasingly accurate as N increases.

In Sect. 3.2, we described how our semiclassical theory reduces to an earlier theory by Schulten and Wolynes under two additional approximations [2]: firstly that the resultant of the nuclear spin vectors is constant, and secondly that the magnitude of this resultant vector has a Gaussian distribution. Both of these approximations become exact in the limit that $N \to \infty$. However, away from that limit the first approximation violates Newton's third law, since there is no torque exerted on the nuclear spins by the electron spin, despite the electron spin experiencing a torque from the nuclear spins. As a result, Schulten and Wolynes' theory is not usually accurate for radical pairs which contain fewer than 100 nuclear spins. However, with these approximations the complexity of the problems is reduced even further – at most, a system of 16 coupled equations must be solved, and in the case of symmetric recombination rates SW theory gives closed form expressions for the singlet and triplet probabilities of radical pairs.

7.1.2 Applications

In Chap. 4, we used the stochastic quantum mechanical method to simulate $PTZ^{\bullet+}$–Ph_n–$PDI^{\bullet-}$ molecular wires of varying lengths, which have a donor-bridge-acceptor structure. There are two possible mechanisms by which charge recombination along this type of molecular wire can occur: the super-exchange mechanism, in which recombination occurs in a single step, or incoherent hopping, in which an electron hops from the bridge to the cation to form an intermediate radical pair, followed by a second electron hopping from the anion to the bridge. Experimental measurements of the overall recombination rate of the $PTZ^{\bullet+}$–Ph_n–$PDI^{\bullet-}$ molecular wires have recently been made [19], which suggested that there may be a change in the mechanism of charge recombination as the length of the wire increases. Unfortunately, from these experiments alone it is not possible to determine the relative contributions of the singlet and triplet recombination pathways to the overall recombination rate.

By comparing the experiments of Weiss et al. to the results of our quantum mechanical spin dynamics calculations, we were able to extract rate constants for the recombination of the singlet and triplet radical pairs. These in turn shed light on the recombination mechanisms of each pathway. We found that recombination of the triplet radical pairs is dominated by the super-exchange mechanism, while the singlet radical pairs recombine by incoherent hopping in all but the shortest molecular wires. The difference in rates can be explained using Marcus theory [20]: direct singlet recombination lies deep in the Marcus inverted region, strongly disfavouring the superexchange mechanism. However, the triplet product is higher in energy than the singlet product, and as a result recombination of the the triplet radical pair by superexchange is far more favourable.

There are two other interesting features of our simulations of the $PTZ^{\bullet+}$–Ph_n–$PDI^{\bullet-}$ molecular wires. Firstly, the very broad resonance peaks seen in the magnetic field effect on the triplet yield of the shorter molecular wires are likely to be the result of lifetime broadening. While the lifetime of the singlet radical pair is reasonably long, the triplet radical pair in these short wires recombines rapidly, which leads to significant broadening of its energy levels. As a result there is a non-zero density of triplet states at the energy of the singlet state over a wide range of magnetic field strengths around the resonance peak, increasing the efficiency of intersystem crossing and hence altering the triplet yield in that range. Secondly, we were unable to even qualitatively explain the experimental magnetic field effect on the triplet yield without introducing a field independent background contribution to the triplet yield, produced by some mechanism outside the normal radical pair reaction scheme. In Sect. 4.4.3 we discussed four possible sources of this background contribution, but none of them appear particularly plausible. Therefore, its origin remains an open question.

In Chap. 5 we applied the semiclassical theory to the problem of avian magnetoreception. We began by reproducing the experimental results of Maeda et al. which demonstrated that an Earth-strength magnetic field could influence the recombination of a carotenoid-porphyrin-fullerene radical pair. By including relaxation in a phenomenological way, we were able to achieve quantitative agreement with experiment at two different magnetic field strengths. Furthermore, we were able to examine the effect of electron spin relaxation on the spin dynamics of the radical pair, and found that both the rate and the origin of the relaxation play an important role.

We then turned our attention to the cryptochrome-based radical pair thought to be responsible for magnetoreception in European robins. We found that the anisotropy in the singlet yield of the [$FAD^{\bullet-}$ $Trp_C^{\bullet+}$] recombination reaction was extremely small. Indeed, we found an anisotropy far smaller than all previous estimates for this radical pair [9, 21]. This difference is due to the fact that, by using the semiclassical theory, we were able to include the electron spin coupling and all of the hyperfine interactions in the radical pair in our simulations, whereas previous quantum mechanical calculations have had to use simplified models of [$FAD^{\bullet-}$ $Trp_C^{\bullet+}$]. The absence of a strong anisotropic response to an Earth-strength magnetic field casts doubt on the plausibility of the [$FAD^{\bullet-}$ $Trp_C^{\bullet+}$] radical pair acting as a magentoreceptor, although recent theoretical work has suggested that a much larger anisotropy in its singlet yield is exhibited if the radical pair is longer-lived than has been previously thought.

Finally, in Chap. 6 we considered the spin dynamics of the polaron pairs found in the semiconducting polymer layer of organic light emitting diodes. These devices exhibit both magnetoelectroluminescence (MEL), a change in the electroluminescence intensity upon application of a magnetic field, and magnetoconductance (MC), a magnetic field dependence of their conductance. We derived a relationship between these two properties and the singlet yield of the polaron pair recombination reaction, and used it to infer the magnetic field effect on the singlet yield (MSY) from the experimentally measured MEL and MC.

Examining the singlet yield of the polaron pair reaction in undeuterated and deuterated DOO-PPV revealed that the hyperfine interaction alone could not account for the magnetic field dependence observed. We proposed that singlet-triplet dephasing, which is the result of the modulation of the exchange interaction between the two polarons, may also play a significant role in the spin dynamics of the polaron pairs. When this dephasing was included in our Schulten-Wolynes calculations of the singlet yield of the polaron pairs in DOO-PPV, the results showed good agreement with experiment, apart from at the very lowest magnetic field strengths. The results also indicated that although it seems likely that the polaron pair mechanism

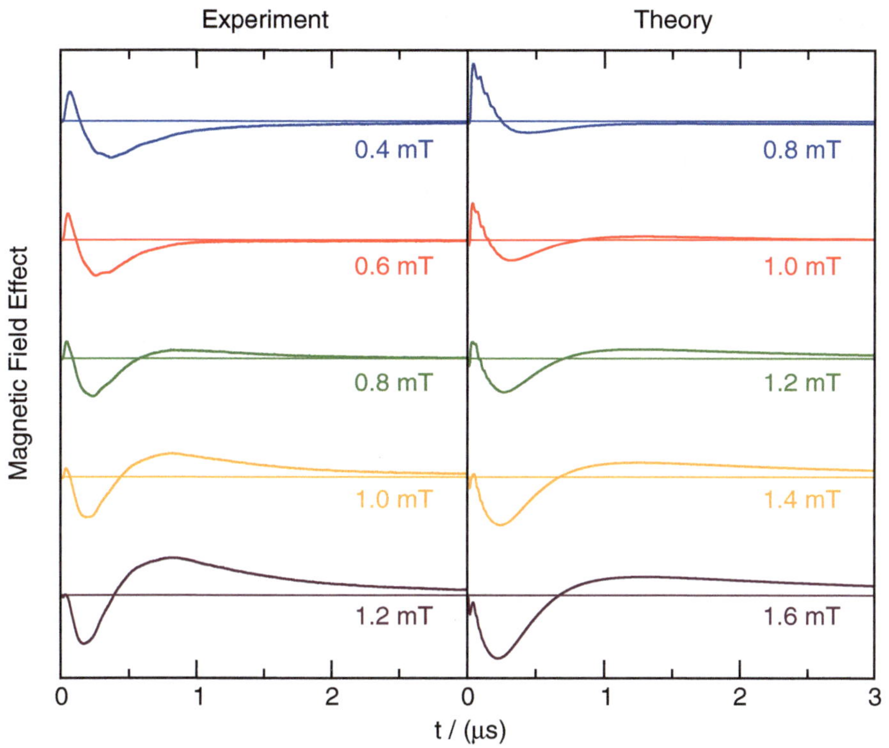

Fig. 7.1 On the left, the magnetic field effect on the experimentally measured absorption of the carotenoid-porphyrin-fullerene radical pair introduced in Sect. 5.2 at a range of applied magnetic field strengths, and on the right the field effect on the simulated survival probability of the radical pair. Experimental data taken from Ref. [22]

plays a significant role in the magnetoconductance of DOO-PPV, it cannot be solely responsible for the observed magnetic field dependence.

7.2 Further Work

7.2.1 Triphasic Magnetic Field Effects

In Sect. 5.2.2 we discussed the magnetic field effect on the transient absorption of the carotenoid-porphyrin-fullerene radical pair, and the biphasic behaviour it exhibits. More recent experiments [22] have shown that at certain magnetic field strengths triphasic behaviour of the magnetic field effect is observed, as shown on the left of Fig. 7.1. Using the semiclassical theory, we have been able to qualitatively reproduce this effect on the simulated survival probability of the radical pair, shown on the

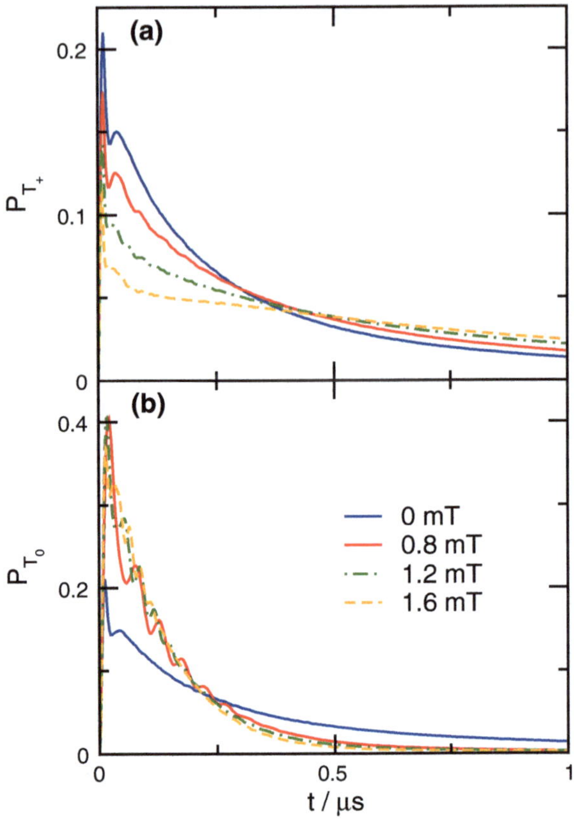

Fig. 7.2 The probability of finding the carotenoid-porphyrin-fullerene radical pair in **a** the $|T_+\rangle$ state and **b** the $|T_0\rangle$ state as a function of time, at a range of applied magnetic field strengths

right of Fig. 7.1, albeit at slightly different magnetic field strengths. This difference is primarily because we have neglected relaxation in these simulations, unlike in Sect. 5.2. There are two reasons for this. Firstly, the relaxation rate will depend on the applied field strength in way that is very hard to predict. As a result, we would need to introduce two free parameters at every magnetic field strength to describe the relaxation, which would not be reasonable. Secondly, we would like to understand the origin of this triphasic behaviour, and interpreting the results produced by the semiclassical theory becomes very difficult when relaxation is accounted for phenomenologically, as discussed in Sect. 5.2.2. Therefore, we have not included it in our simulations.

We found that the triphasic behaviour can be explained by examining the how the probability of finding the radical pair in the $|T_0\rangle$, $|T_+\rangle$, and $|T_-\rangle$ states changes with the magnetic field strength. The magnetic field dependence of the probability of being in the $|T_+\rangle$ state, P_{T+}, is shown in Fig. 7.2a, and is nearly identical to P_{T-}, not shown. This is relatively straightforward to understand – it is simply the high field effect described in Sect. 1.3.1. As the magnetic field strength increases, the $|T_+\rangle$ and $|T_-\rangle$ states become increasingly separated in energy from the $|S\rangle$ state the radical pair is formed in, and so the rate of $|S\rangle \rightarrow |T_+\rangle$ and $|S\rangle \rightarrow |T_-\rangle$ interconversion is reduced. As a result, the height of the maximum in P_{T+} and P_{T-} seen at short times decreases as the strength of the magnetic field increases, while at later times P_{T+} and P_{T-} increase with the magnetic field strength, because the backward $|T_+\rangle \rightarrow |S\rangle$ and $|T_-\rangle \rightarrow |S\rangle$ transitions become less likely.

By contrast, Fig. 7.2b shows that the probability of being found in the $|T_0\rangle$ state at short times is larger when a field is applied than in the absence of a field, although the strength of the applied field does not seem to have an effect. This is a manifestation of the low field effect described in Sect. 1.3.2. Applying a magnetic field does not affect the energy of the $|T_0\rangle$ state, but does break the symmetry of the system and allow more paths for interconversion between the $|S\rangle$ and $|T_0\rangle$ states. This also results in a smaller P_{T0} in the presence of a magnetic field at longer times, since $|T_0\rangle \rightarrow |S\rangle$ back conversion is more likely.

Together, these observations explain the triphasic behaviour observed in Fig. 7.1. At low fields, the LFE in P_{T0} dominates the survival probability, which is larger than the survival probability in the absence of a field at short times and smaller at longer times. The reverse is true at high fields, when the HFE in P_{T+} and P_{T-} dominates. Since the application of a magnetic field affects the different triplet states in different ways, at intermediate fields there is competition between the two effects, resulting in the triphasic behaviour observed. While including relaxation in our semiclassical theory will certainly complicate this picture, we believe that even when relaxation is accounted for this qualitative description will remain the same.

The semiclassical theory has given us a deep mechanistic insight into the spin dynamics of this radical pair, and the origin of the low and high field effects. However, the question of why the low field effect promotes $|S\rangle \leftrightarrow |T_0\rangle$ interconversion and not conversion between the $|S\rangle$ and $|T_\pm\rangle$ states still requires some further analysis before this work can be written up for publication.[2]

7.2.2 *Relaxation*

The relaxation of the electron spins in a radical pair is extremely complicated to describe, as it involves the coupling of the spin system to the nuclear motions of the radical pair. The reasons for the difficulty in accounting for this system-environment coupling in quantum mechanical simulations are outlined in Sect. 2.7.1. In the semi-classical theory, it is relatively straightforward to include electron spin relaxation in a crude, phenomenological way, as shown in Sect. 3.4. However, this approach may not produce accurate results, and provides no information about the microscopic origin of relaxation.

A newly developed method by Lindoy and Manolopoulos has aimed to incorporate the molecular motions of the radical pair directly into spin dynamics calculations

[2]Since this thesis was written this work has been completed and published in *Journal of Chemical Physics* **149**, 034103 (2018).

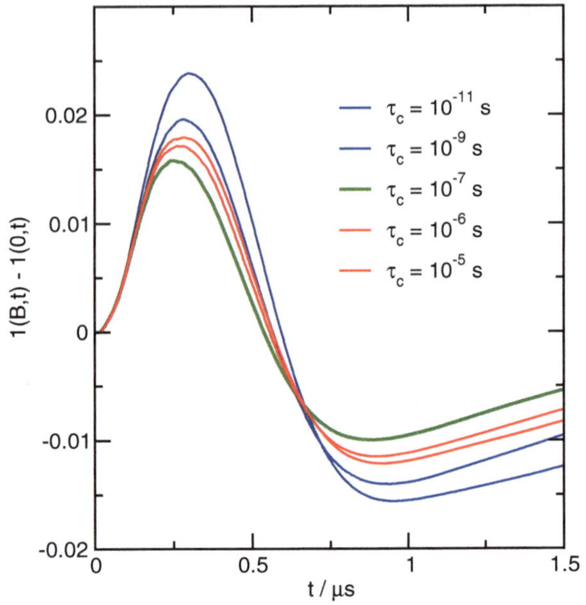

Fig. 7.3 The magnetic field effect on the survival probability of the carotenoid-porphyrin-fullerene radical pair introduced in Sect. 5.2 tumbling around the long axis of the carotenoid with a range of different correlation times, τ_C. Data reproduced with permission from Ref. [24]

by explicitly simulating the nuclear motion of the radical pair and re-evaluating the interactions which contribute to the spin Hamiltonian after each step [23, 24]. For example, since rotating the radical pair changes the anisotropic dipolar components of the hyperfine tensors, rotational motion modulates these hyperfine tensors, causing the spin Hamiltonian to become time-dependent and inducing relaxation of the electron spins. This approach is extremely general: so far, the nuclear motion has only been treated using a simple classical stochastic model [23, 24], but in principle more sophisticated simulations of the nuclear dynamics could be coupled to the spin dynamics.

This approach has been used to try to account for the relaxation of the carotenoid-porphyrin-fullerene radical pair introduced in Sect. 5.2 by coupling the spin dynamics of the radical pair to its rotational motion around the long axis of the carotenoid [24]. The effect of a 49 μT magnetic field on the survival probability of a $C^{\bullet+}PF^{\bullet-}$ radical pair tumbling with a range of rotational correlation times, τ_C, is shown in Fig. 7.3. The signal produced in the fast tumbling limit (corresponding to a small τ_C) differs significantly from that observed in the slow tumbling limit (large τ_C), and the most efficient relaxation is observed at some intermediate tumbling rate, shown in green. While including this source of relaxation does improve the agreement with the experimental results shown in Fig. 5.4, accounting for the rotation around the long axis alone is not sufficient to quantitatively reproduce the experiments. Clearly a more detailed description of the nuclear motion is required.

It is clearer to see that coupling the rotational motion to the spin dynamics does indeed result in the relaxation of the electron spins in Fig. 7.4, which shows the probability of finding the carotenoid-porphyrin-fullerene radical pair in the singlet

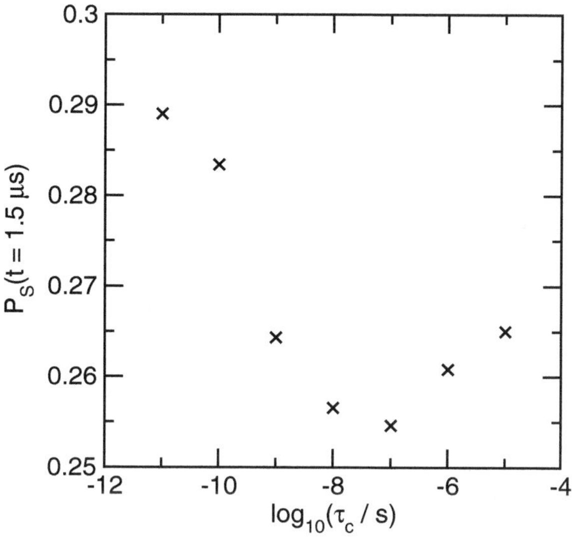

Fig. 7.4 The singlet probability of the carotenoid-porphyrin-fullerene radical pair tumbling around the long axis of the carotenoid after 1.5 µs as a function of the rotational correlation time. The equilibrium value of P_S is 1/4. In all cases the statistical errors are smaller than the sizes of the plotted points. Data reproduced with permission from Ref. [9]

state 1.5 µs after it is formed, in the absence of either an applied magnetic field or any recombination of the radical pair. The effect of relaxation is always to cause the singlet probability to tend towards its equilibrium value of 1/4. Therefore, the closer the singlet probability is to this value after a certain amount of time, the more efficiently relaxation is occurring. In this case, Fig. 7.4 shows that relaxation is fastest when the rotational correlation time τ_C is between 10^{-7} s and 10^{-8} s^{-1}, and becomes less efficient in both the slow and fast tumbling limit. This type of behaviour is expected when the rotational motion is Brownian, in which case the relaxation rate is proportional to [25]

$$k_R \propto \frac{2\tau_C}{1+\omega_t^2\tau_C^2}, \tag{7.1}$$

where ω_t is the transition frequency of the electron spin.

While this technique is still in its infancy, it has great potential. Firstly, rather than coupling the molecular motions of the radical pair to a semiclassical simulation of their spin dynamics, as has been done in this example, it is also possible to use an exact quantum mechanical method in a Hilbert space formulation to treat the spin dynamics of the radical pair, such as the method described in Sect. 2.5. This approach to accounting for relaxation is far more efficient than existing quantum mechanical techniques, which require calculations to be performed in Liouville space. Secondly, there are significant advantages to describing relaxation by explicitly considering the molecular motion of the radical pair, as information about these motions is typically easier to obtain than the relaxation times themselves, and unlike T_1 and T_2 nuclear motions are independent of the applied magnetic field. Alternatively, by taking the parameters which describe the motion of the radical pair to be free and then fitting the simulated results to experimental data, it may be possible to use this technique to deduce information about the molecular motions of a radical pair from its spin dynamics.

References

1. Kaptein, R., & Oosterhoff, J. (1969). Chemically induced dynamic nuclear polarization II. *Chemical Physics Letters, 4*, 195–197.
2. Schulten, K., & Wolynes, P. G. (1978). Semi-classical description of electron-spin motion in radicals including effect of electron hopping. *Journal of Chemical Physics, 68*, 3292–3297.
3. McLauchlan, K. A., & Steiner, U. E. (1991). The spin-correlated radical pair as a reaction intermediate. *Molecular Physics, 73*, 241–263.
4. Ehrenfreund, E., & Vardeny, Z. V. (2012). Effects of magnetic field on conductance and electroluminescence in organic devices. *Israel Journal of Chemistry, 52*, 552–562.
5. Hore, P. J., & Mouritsen, H. (2016). The radical-pair mechanism of magnetoreception. *Annual Review of Biophysics, 45*, 299.
6. Brocklehurst, B., & McLauchlan, K. A. (1996). Free radical mechanism for the effects of environmental electromagnetic fields on biological systems. *International Journal of Radiation Biology, 69*, 3–24.
7. Wasielewski, M. R. (2006). Energy, charge, and spin transport in molecules and self-assembled nanostructures inspired by photosynthesis. *Journal of Organic Chemistry, 71*, 5051–5066.
8. Ritz, T., Adem, S., & Schulten, K. (2000). A model for photoreceptor-based magnetoreception in birds. *Biophysical Journal, 78*, 707–718.
9. Rodgers, C. T., & Hore, P. J. (2009). Chemical magnetoreception in birds: The radical pair mechanism. *Proceedings of the National Academy of Sciences of the United States of America, 106*, 353–360.
10. Tang, C. W., & Vanslyke, S. A. (1987). Organic electroluminescent diodes. *Applied Physics Letters, 51*, 913–915.
11. Kido, J. (1999). Organic displays. *Physics World, 12*, 27–30.
12. Brocklehurst, B. (1976). Magnetic field effect on the pulse shape of scintillations due to geminate recombination of ion pairs. *Chemical Physics Letters, 44*, 245–248.
13. Weller, A., Nolting, F., & Staerk, H. (1983). A quantitative interpretation of the magnetic field effect on hyperfine-coupling-induced triplet fromation from radical ion pairs. *Chemical Physics Letters, 96*, 24–27.
14. Weller, A., Staerk, H., & Treichel, R. (1984). Magnetic-field effects on geminate radical-pair recombination. *Faraday Discussions of the Chemical Society, 78*, 271.
15. Steiner, U. E., & Ulrich, T. (1989). Magnetic field effects in chemical kinetics and related phenomena. *Chemical Reviews, 89*, 51–147.
16. Till, U., & Hore, P. J. (1997). Radical pair kinetics in a magnetic field. *Molecular Physics, 90*, 289–296.
17. Timmel, C. R., Till, U., Brocklehurst, B., Mclauchlan, K. A., & Hore, P. J. (1998). Effects of weak magnetic fields on free radical recombination reactions. *Molecular Physics, 95*, 71–89.
18. Till, U., Timmel, C. R., Brocklehurst, B., & Hore, P. J. (1998). The influence of very small magnetic fields on radical recombination reactions in the limit of slow recombination. *Chemical Physics Letters, 298*, 7–14.
19. Weiss, E. A., et al. (2004). Making a molecular wire: Charge and spin transport through. *Journal of the American Chemical Society, 126*, 5577–5584.
20. Marcus, R. A. (1965). On the theory of electron-transfer reactions. VI. Unified treatment for homogeneous and electrode reactions. *Journal of Chemical Physics, 43*, 679.
21. Lee, A. A., et al. (2014). Alternative radical pairs for cryptochrome-based magnetoreception. *Journal of the Royal Society Interface, 11*, 20131063.

22. West, A. (2017). *Magnetic field effects in carotenoid-porphyrin-fullerene triad molecules part II*. Thesis: University of Oxford.
23. Lindoy, L. P. J., & Manolopoulos, D. E. (2017). Electron spin relaxation in the Hilbert space formulation of radial pair recombination reactions. Unpublished manuscript.
24. Vanderstichele, B. (2017). *Electron spin relaxation in the semiclassical theory of radical pair reactions part II*. Thesis: University of Oxford.
25. Carrington, A., & McLachlan, A. D. (1967). *Introduction to magnetic resonance with applications to chemistry and chemical physics*. New York: Harper and Row.

Appendix A
Wavepacket Propagation Techniques

Both the deterministic and stochastic quantum mechanical methods of evaluating the ensemble average of an observable described in Chap. 2 involve propagating a number of wavefunctions to time t. This propagation is described by

$$|\Psi; t\rangle = e^{-i\hat{H}t-\hat{K}t}|\Psi; 0\rangle, \tag{A.1}$$

where $\hat{H}$ is the spin Hamiltonian and $\hat{K}$ is the recombination operator. While there exist a range of sophisticated techniques designed for solving equations of this form, they must be slightly modified for use here because the operator $\left(\hat{H} - i\hat{K}\right)$ is not Hermitian. We will now briefly describe two methods which can be used to carry out this propagation.

A.1 The Short Iterative Arnoldi Method

The Short Iterative Arnoldi (SIA) method [1, 2] is a generalisation of the Short Iterative Lanczos method of solving the Schrödinger equation [3, 4] which allows for non-Hermitian effective Hamiltonians $\hat{\mathcal{H}}$. It is based on the Taylor series expansion of the propagator,

$$e^{-i\hat{\mathcal{H}}\delta t} = \sum_{k=0}^{\infty} \frac{1}{k!}\left(-i\delta t\hat{\mathcal{H}}\right)^k. \tag{A.2}$$

By truncating this expansion when $k = p - 1$, we may write the wavefunction at time $t + \delta t$ as

$$|\Psi; t+\delta t\rangle = \sum_{k=0}^{p-1} \frac{1}{k!}\left(-i\delta t\hat{\mathcal{H}}\right)^k |\Psi; t\rangle + \mathcal{O}\left((\delta t\hat{\mathcal{H}})^p\right)|\Psi; t\rangle. \tag{A.3}$$

A. Lewis, *Spin Dynamics in Radical Pairs*, Springer Theses,
https://doi.org/10.1007/978-3-030-00686-0

Therefore, within an error of $\mathcal{O}(\delta t^p)$, $|\Psi; t+\delta t\rangle$ may be written as linear combination of wavefunctions which form a p dimensional Hilbert space called the Krylov subspace,

$$\begin{aligned}\mathcal{K}_p &\equiv \left\{|\Psi;t\rangle, \hat{\mathcal{H}}|\Psi;t\rangle, \cdots, \hat{\mathcal{H}}^{p-1}, |\Psi;t\rangle\right\} \\ &\equiv \left\{|q_0\rangle, |q_1\rangle, \cdots, |q_{p-1}\rangle\right\},\end{aligned} \tag{A.4}$$

with time dependent coefficients $c_k(t)$:

$$|\Psi; t+\delta t\rangle \simeq \sum_{k=0}^{p-1} c_k(t)|q_k\rangle. \tag{A.5}$$

In order to calculate the coefficients $c_k(t)$, we must find the matrix representation $\mathbf{A}$ of the effective Hamiltonian $\hat{\mathcal{H}}$ in the basis of orthogonal Krylov vectors, whose elements are defined by

$$A_{j,k} = \langle q_j|\hat{\mathcal{H}}|q_k\rangle. \tag{A.6}$$

The Arnoldi algorithm outlined in Algorithm 1 is used to construct both the Krylov subspace $\mathcal{K}_p$ and the representation of the effective Hamiltonian in that basis, $\mathbf{A}$. The time propagation during the interval $\tau < \delta t$ is then described by

$$\mathbf{c}(t+\tau) = e^{-i\mathbf{A}\tau}\mathbf{c}(t), \tag{A.7}$$

where $\mathbf{c}(t)$ is the p-dimensional vector of the coefficients $c_k(t)$. Since p is chosen to be far smaller than the size of the Hilbert space, the propagation between times t and $t+\delta t$ is inexpensive.

Algorithm 1 Arnoldi's algorithm, adapted from Ref. [2].

$|q_0\rangle \leftarrow |\Psi;t\rangle/\sqrt{\langle\Psi;t|\Psi;t\rangle}$
for $k = 0$ to $p-1$, **do**
  $|p\rangle \leftarrow \hat{\mathcal{H}}|q_k\rangle$
  for $j = 0$ to k, **do**
    $A_{j,k} \leftarrow \langle q_k|p\rangle$
    $|p\rangle \leftarrow |p\rangle - A_{j,k}|q_j\rangle$
  end for
  if $k < p-1$ **then**
    $A_{k+1,k} \leftarrow \sqrt{\langle p|p\rangle}$
    $|q_{k+1}\rangle \leftarrow |p\rangle/A_{k+1,k}$
  end if
end for

This approximation to the wavefunction will remain accurate provided that the action of $\hat{\mathcal{H}}$ on $|\Psi; t+\tau\rangle$ is accurately described by a linear combination of the Krylov states. Since $\hat{\mathcal{H}}|q_k\rangle$ is another Krylov state for all k except $k = p-1$, this condition will be satisfied as long as $c_{p-1}(t)$ is small compared to the norm of the vector of coefficients, $|\mathbf{c}|$. Once $c_{p-1}(t)/|\mathbf{c}|$ reaches some cut-off ϵ, a new Krylov subspace is generated and the processes is restarted. Hence by reducing ϵ, this method allows the propagation of wavefunctions under the effective Hamiltonian $\hat{\mathcal{H}} = \left(\hat{H} - i\hat{K}\right)$ with arbitrary precision.

A.2 The Split Operator/Symplectic Integrator Method

The split operator approach is based on an alternative expansion of the propagator. In this case, the propagator is approximately factorised into a product of exponentials [5],

$$e^{-i\hat{H}\delta t - w\hat{K}\delta t} \approx e^{-\hat{K}\delta t/2} e^{-i\hat{H}\delta t} e^{-\hat{K}\delta t/2}, \tag{A.8}$$

where $e^{-i\hat{H}\delta t}$ is a unitary, Hermitian propagator which describes the spin evolution of the radical pair, and $e^{-\hat{K}\delta t/2}$ is a non-unitary propagator which describes the recombination of the radical pair. Since $\hat{H}$ and $\hat{K}$ do not in general commute, this approximation is only accurate to $\mathcal{O}(\delta t^3)$. The action of the two separate propagators on a wavefunction is much more straightforward to evaluate than that of the exact propagator, as we shall now demonstrate.

The Haberkorn recombination operator $\hat{K}$ is defined in Sect. 2.2.1 as [6]

$$\hat{K} = \frac{k_{\mathrm{S}}}{2}\hat{P}_{\mathrm{S}} + \frac{k_{\mathrm{T}}}{2}\hat{P}_{\mathrm{T}}. \tag{A.9}$$

Using this definition of $\hat{K}$, we may write

$$e^{-\hat{K}\delta t/2} = e^{-k_{\mathrm{S}}\hat{P}_{\mathrm{S}}\delta t/4} e^{-k_{\mathrm{T}}\hat{P}_{\mathrm{T}}\delta t/4}, \tag{A.10}$$

since $\hat{P}_{\mathrm{S}} + \hat{P}_{\mathrm{T}} = \hat{1}$ and so $\hat{P}_{\mathrm{S}}$ and $\hat{P}_{\mathrm{T}}$ commute. A projection operator must be idempotent ($\hat{P}^2 = \hat{P}$)[7], and so the exponential of a projection operator may be expressed as

$$\begin{aligned} e^{\beta\hat{P}} &= \sum_{k=0}^{\infty} \frac{(\beta\hat{P})^k}{k!} \\ &= \hat{1} + \hat{P}\sum_{k=1}^{\infty} \frac{\beta^k}{k!} \\ &= \hat{1} + \hat{P}\left(e^{\beta} - 1\right) \quad = \hat{P}e^{\beta} + \left(\hat{1} - \hat{P}\right). \end{aligned} \tag{A.11}$$

Using these results, and the fact that $\hat{P}_{\mathrm{S}}\hat{P}_{\mathrm{T}} = \hat{P}_{\mathrm{S}}(\hat{1} - \hat{P}_{\mathrm{S}}) = 0$, the non-unitary propagator becomes

$$\begin{aligned}
e^{-\hat{K}\delta t/2} &= e^{-k_\mathrm{S}\hat{P}_\mathrm{S}\delta t/4} e^{-k_\mathrm{T}\hat{P}_\mathrm{T}\delta t/4} \\
&= \left[\hat{P}_\mathrm{S} e^{-k_\mathrm{S}\delta t/4} + \left(\hat{1} - \hat{P}_\mathrm{S}\right)\right]\left[\hat{P}_\mathrm{T} e^{-k_\mathrm{T}\delta t/4} + \left(\hat{1} - \hat{P}_\mathrm{T}\right)\right] \\
&= \left[\hat{P}_\mathrm{S} e^{-k_\mathrm{S}\delta t/4} + \hat{P}_\mathrm{T}\right]\left[\hat{P}_\mathrm{T} e^{-k_\mathrm{T}\delta t/4} + \hat{P}_\mathrm{S}\right] \\
&= \hat{P}_\mathrm{S}\hat{P}_\mathrm{T}\left(e^{-(k_\mathrm{S}+k_\mathrm{T})\delta t/4} + 1\right) + \hat{P}_\mathrm{S}^2 e^{-k_\mathrm{S}\delta t/4} + \hat{P}_\mathrm{T}^2 e^{-k_\mathrm{T}\delta t/4} \\
&= \hat{P}_\mathrm{S}\, e^{-k_\mathrm{S}\delta t/4} + \hat{P}_\mathrm{T}\, e^{-k_\mathrm{T}\delta t/4}.
\end{aligned} \tag{A.12}$$

The projection operators $\hat{P}_\mathrm{S}$ and $\hat{P}_\mathrm{T}$ are diagonal in the coupled basis of the electron spins used in Chap. 2, and so the action of the non-unitary propagator $e^{-\hat{K}\delta t/2}$ on a state represented in this basis is straightforward to calculate exactly.

Calculating the action of the unitary propagator $e^{-i\hat{H}\delta t}$ on a state $|\Psi; t\rangle$ is equivalent to solving the Schrödinger equation,

$$i\frac{\mathrm{d}}{\mathrm{d}t}|\Psi; t\rangle = \hat{H}|\Psi; t\rangle. \tag{A.13}$$

We chose to use a symplectic integrator to do this [8-11]. In this approach, the wavefunction is expanded in a basis $\{|\phi_k\rangle\}$ in which the Hamiltonian is real and symmetric,[1]

$$|\Psi; t\rangle = \sum_k a_k(t)|\phi_k\rangle. \tag{A.14}$$

$a_k(t)$ is a complex time dependent coefficient, which may be written in terms of its real and imaginary parts, $a_k(t) = q_k(t) + ip_k(t)$. We may define

$$\begin{aligned}
|p; t\rangle &= \sum_k p_k(t)|\phi_k\rangle, \\
|q; t\rangle &= \sum_k q_k(t)|\phi_k\rangle,
\end{aligned} \tag{A.15}$$

so that

$$|\Phi; t\rangle = |q; t\rangle + i|p; t\rangle. \tag{A.16}$$

Inserting this definition of $|\Phi; t\rangle$ into the Schrödinger equation in Eq. (A.13) and equating the complex and real coefficients gives

[1] It is also possible to treat complex and time-dependent Hamiltonians using a symplectic integrator with a modified version of the following argument, as shown in Ref. [8].

$$
\begin{aligned}
\frac{\mathrm{d}}{\mathrm{d}t} q_k(t) &= \sum_j H_{kj}\, p_j(t), \\
\frac{\mathrm{d}}{\mathrm{d}t} p_k(t) &= -\sum_j H_{kj}\, q_j(t),
\end{aligned}
\tag{A.17}
$$

where $H_{kj} = \langle \phi_k | \hat{H} | \phi_j \rangle$. Defining a classical Hamiltonian

$$
H_C(t) = \sum_{j,k} H_{jk} \left(p_j(t) p_k(t) + q_j(t) q_k(t) \right), \tag{A.18}
$$

the equations in Eq. (A.17) can be rewritten as Hamilton's classical equations of motion,

$$
\begin{aligned}
\frac{\mathrm{d}}{\mathrm{d}t} q_k(t) &= \frac{\partial H_C}{\partial p_k}, \\
\frac{\mathrm{d}}{\mathrm{d}t} p_k(t) &= -\frac{\partial H_C}{\partial q_k},
\end{aligned}
\tag{A.19}
$$

These classical equations of motion may be solved by an mth order symplectic integration scheme, given in Algorithm 2, which is accurate to $\mathcal{O}(\delta t^m)$. Hence by splitting the propagator into a unitary and non-unitary part and using a symplectic integrator to perform the unitary propagation of the wavefunction, we can efficiently and accurately solve Eq. (A.1).

Algorithm 2 A Symplectic Integrator algorithm, adapted from Ref. [8]. The real coefficients a_j and b_j are given in Ref. [8] for integrators of various different values of m.

$|q\rangle \leftarrow |q; t\rangle$
$|p\rangle \leftarrow |p; t\rangle$
for $j = 1$ to m, **do**
  $|p\rangle \leftarrow |p\rangle - b_j\, \delta t \hat{H} |q\rangle$
  $|q\rangle \leftarrow |q\rangle + a_j\, \delta t \hat{H} |p\rangle$
end for
$|q; t + \delta t\rangle \leftarrow |q\rangle$
$|p; t + \delta t\rangle \leftarrow |p\rangle$

References

1. Saad, Y. (1992). Analysis of some Krylov subspace approximations to the matrix exponential operator. *SIAM Journal on Numerical Analysis 29*, 209–228.
2. Pollard, W. T. & Friesner, R. A. (1994). Solution of the Redfield equation for the dissipative quantum dynamics of multilevel systems. *The Journal of Chemical Physics 100*, 5054–5065 (1994).
3. Park, T. J. & Light, J. C. (1986). Unitary quantum time evolution by iterative Lanczos reduction. *Journal of Chemical Physics 85*, 5870.

4. Leforestier, C. et al. (1991). A comparison of different propagation schemes for the time dependent Schrödinger equation. *Journal of Computational Physics 94*, 59–80.
5. Feit, M. D., Fleck, J. A. & Steiger, A. (1982). Solution of the Schroedinger equation by a spectral method. *Journal of Computational Physics 47*, 412–433.
6. Haberkorn, R. (1976). Density matrix description of spin-selective radical pair reactions. *Molecular Physics 32*, 1491–1493.
7. Trigg, G. (1964). *Quantum mechanics*. Van Nostrand.
8. Gray, S. K. & Manolopoulos, D. E. (1996). Symplectic integrators tailored to the time-dependent Schrödinger equation. *The Journal of Chemical Physics 104*, 7099–7112 (1996).
9. Ruth, R. D. (1983). A canonical integration technique. *IEEE Transactions on Nuclear Science*, 2228–2230.
10. Gray, S. K. & Verosky, J. M. (1994). Classical Hamiltonian structures in wave packet dynamics. *The Journal of Chemical Physics 100*, 5011–5022.
11. Hoover, W. G., Kum, O. & Owens, N. E. (1995). Accurate symplectic integrators via random sampling. *The Journal of Chemical Physics 103*, 1530–1532.

Appendix B
Rotationally Averaged Dipolar Coupling

In order to see the effect of tumbling in solution upon the dipolar coupling, it is helpful to write out the coupling matrix **D** explicitly in spherical coordinates:

$$\mathbf{D} = D \begin{pmatrix} 1 - 3\sin^2\theta\cos^2\phi & 3\sin^2\theta\cos\phi\sin\phi & 3\sin\theta\cos\theta\cos\phi \\ 3\sin^2\theta\cos\phi\sin\phi & 1 - 3\sin^2\theta\sin^2\phi & 3\sin\theta\cos\theta\sin\phi \\ 3\sin\theta\cos\theta\cos\phi & 3\sin\theta\cos\theta\sin\phi & 1 - 3\cos^2\theta \end{pmatrix}. \tag{B.1}$$

To average over all relative orientations of the two electrons spins, we must evaluate

$$\int_0^{\pi} \sin\theta \mathrm{d}\theta \int_0^{2\pi} \mathrm{d}\phi\, \mathbf{D} = \left(\int_0^{\pi} \sin\theta \mathrm{d}\theta \int_0^{2\pi} \mathrm{d}\phi\, d_{ij} \right). \tag{B.2}$$

Noting that

$$\int_0^{2\pi} \sin\phi\, \mathrm{d}\phi = \int_0^{2\pi} \cos\phi\, \mathrm{d}\phi = \int_0^{2\pi} \sin\phi\cos\phi\, \mathrm{d}\phi = 0, \tag{B.3}$$

it is immediately apparent that the off-diagonal elements of this matrix equal zero. Then, using the results

$$\begin{gathered} \int_0^{\pi} \sin\theta\, \mathrm{d}\theta = 2, \int_0^{\pi} \sin^2\theta\, \mathrm{d}\theta = \frac{4}{3}, \int_0^{\pi} \cos^2\theta\, \mathrm{d}\theta = \frac{2}{3} \\ \int_0^{2\pi} \mathrm{d}\phi\ = 2\pi, \int_0^{2\pi} \sin^2\phi\, \mathrm{d}\phi = \int_0^{2\pi} \cos^2\phi\, \mathrm{d}\phi = \pi, \end{gathered} \tag{B.4}$$

the diagonal elements can also be shown to equal zero. Therefore, when the relative orientations of the electrons are changing rapidly, as is the case for a radical pair tumbling in solution, the dipolar coupling averages to zero.

A. Lewis, *Spin Dynamics in Radical Pairs*, Springer Theses,
https://doi.org/10.1007/978-3-030-00686-0

Appendix C
Generalising the Semiclassical Equations of Motion

The coupling between the two electron spins in a radical pair is described by the term $\hat{\mathbf{S}}_1 \cdot \mathbf{D}' \cdot \hat{\mathbf{S}}_2$ in the spin Hamiltonian in Eq. (2.1), where $\mathbf{D}'$ is a tensor which accounts for both the dipolar and exchange coupling between the two electrons, as defined in Sect. 2.1.1. For simplicity, this was neglected when we derived our semiclassical equations of motion in Sect. 3.1. When it is included, the following additional terms appear in the semiclassical equations of motion in Eqs. (3.4)–(3.6):

$$\frac{d}{dt}\mathbf{S}_1 = \cdots + \mathbf{D}' \times \mathbf{T}_{12}, \tag{C.1}$$

$$\frac{d}{dt}\mathbf{S}_2 = \cdots - \mathbf{D}' \times \mathbf{T}_{12}, \tag{C.2}$$

$$\frac{d}{dt}\mathbf{T}_{12} = \cdots - \mathbf{S}_1 \times \mathbf{D}' - \mathbf{D}' \times \mathbf{S}_2. \tag{C.3}$$

In Eqs. (C.1) and (C.2), $\mathbf{D}' \times \mathbf{T}_{12}$ denotes the sum of the vector products of each column of $\mathbf{D}'$ with the corresponding column of $\mathbf{T}_{12}$, and in Eq. (C.3) $\mathbf{S}_1 \times \mathbf{D}'$ denotes the vector product of $\mathbf{S}_1$ with each column of $\mathbf{D}'$, and $\mathbf{D}' \times \mathbf{S}_2$ denotes the vector product of each row of $\mathbf{D}'$ with $\mathbf{S}_2$.

In Sect. 3.1 we also assumed that the hyperfine interactions between the electron and nuclear spins were isotropic. If we remove this assumption, the effective magnetic field about which the electron spin in radical i precesses is

$$\bar{\boldsymbol{\omega}}_i = \boldsymbol{\omega}_i + \sum_{k=1}^{N_i} \mathbf{A}'_{ik} \cdot \mathbf{I}_{ik}, \tag{C.4}$$

where $\mathbf{A}'_{ik}$ is the hyperfine coupling tensor which accounts for both the Fermi contact interaction and the dipolar coupling between the kth nuclear spin and the electron in radical i, as defined in Sect. 2.1.4. The original equations of motion for the nuclear spins in Eq. (3.3), which were derived from the Heisenberg equation of motion, are replaced by

A. Lewis, *Spin Dynamics in Radical Pairs*, Springer Theses,
https://doi.org/10.1007/978-3-030-00686-0

$$\frac{d}{dt}\mathbf{I}_{ik} = \left(\mathbf{A}'_{ik} \cdot \mathbf{S}_i\right) \times \mathbf{I}_{ik} - \bar{k}\,\mathbf{I}_{ik} + 4\Delta k\,\mathrm{tr}[\mathbf{T}_{12}]\mathbf{I}_{ik}, \tag{C.5}$$

and the modified equations of motion for the nuclear spins in Eq. (3.24) become

$$\frac{d}{dt}\mathbf{I}_{ik} = \frac{\sqrt{S_i(S_i+1)}}{|\mathbf{S}_i|}\left(\mathbf{A}'_{ik} \cdot \mathbf{S}_i\right) \times \mathbf{I}_{ik}. \tag{C.6}$$

The equations of motion of the classical spin variables evolving under the full Hamiltonian in Eq. (2.1) can alternatively be written in tensor notation, using the Einstein summation convention for the Greek (Cartesian component) indices:

$$\frac{d}{dt}S^{(1)}_\alpha = \epsilon_{\alpha\beta\gamma}\left(\omega^{(1)}_\beta S^{(1)}_\gamma + D'_{\beta\delta}T_{\gamma\delta}\right) - \bar{k}S^{(1)}_\alpha + \Delta k S^{(2)}_\alpha, \tag{C.7}$$

$$\frac{d}{dt}S^{(2)}_\alpha = \epsilon_{\alpha\beta\gamma}\left(\omega^{(2)}_\beta S^{(2)}_\gamma + D'_{\delta\beta}T_{\delta\gamma}\right) - \bar{k}S^{(2)}_\alpha + \Delta k S^{(1)}_\alpha, \tag{C.8}$$

$$\begin{aligned}\frac{d}{dt}T_{\alpha\beta} = \epsilon_{\alpha\gamma\delta}\left(\omega^{(1)}_\gamma T_{\delta\beta} - \frac{1}{4}S^{(1)}_\gamma D'_{\delta\beta}\right) + \epsilon_{\beta\gamma\delta}\left(\omega^{(2)}_\gamma T_{\alpha\delta} - \frac{1}{4}S^{(2)}_\gamma D'_{\alpha\delta}\right) - \\ \bar{k}T_{\alpha\beta} + \Delta k T_{\beta\alpha} + \Delta k \delta_{\alpha\beta}\left(\frac{1}{4}\bar{\mathbb{1}} - T_{\gamma\gamma}\right),\end{aligned} \tag{C.9}$$

$$\frac{d}{dt}\bar{\mathbb{1}} = -\bar{k}\bar{\mathbb{1}} + 4\Delta k T_{\alpha\alpha}. \tag{C.10}$$

Here $\epsilon_{\alpha\gamma\delta}$ is the alternating tensor, $\delta_{\alpha\beta}$ is the Kronecker delta, $S^{(i)}_\alpha$ is the α component of $\mathbf{S}_i$, $T_{\alpha\beta}$ is the classical variable corresponding to the quantum mechanical operator $\hat{S}^{(1)}_\alpha \hat{S}^{(2)}_\beta$, $\bar{k}$ and Δk are still defined by Eqs. (3.8) and (3.9) respectively, and

$$\omega^{(i)}_\alpha = -\gamma_i B_\alpha + \sum_{k=1}^{N_i} A'^{(ik)}_{\alpha\beta} I^{(ik)}_\beta, \tag{C.11}$$

where γ_i is the gyromagnetic ratio of the electron in radical i, and B_α is the α component of the applied magnetic field **B**. The original equations of motion for the nuclear spins in Eq. (C.5) are written as

$$\frac{d}{dt}I^{(ik)}_\alpha = \epsilon_{\alpha\beta\gamma} A'^{(ik)}_{\beta\delta} S^{(i)}_\delta I^{(ik)}_\gamma + \left(4\Delta k T_{\beta\beta} - \bar{k}\right) I^{(ik)}_\alpha \tag{C.12}$$

in tensor notation, while our modified equations of motion in Eq. (C.6) become

$$\frac{d}{dt}I^{(ik)}_\alpha = \frac{\sqrt{S_i(S_i+1)}}{|\mathbf{S}_i|}\epsilon_{\alpha\beta\gamma} A'^{(ik)}_{\beta\delta} S^{(i)}_\delta I^{(ik)}_\gamma. \tag{C.13}$$

In Sect. 3.1 we showed numerically that the equations of motion in Eqs. (3.4)–(3.7) are exact in the limit of a radical pair with no nuclear spins. This remains true

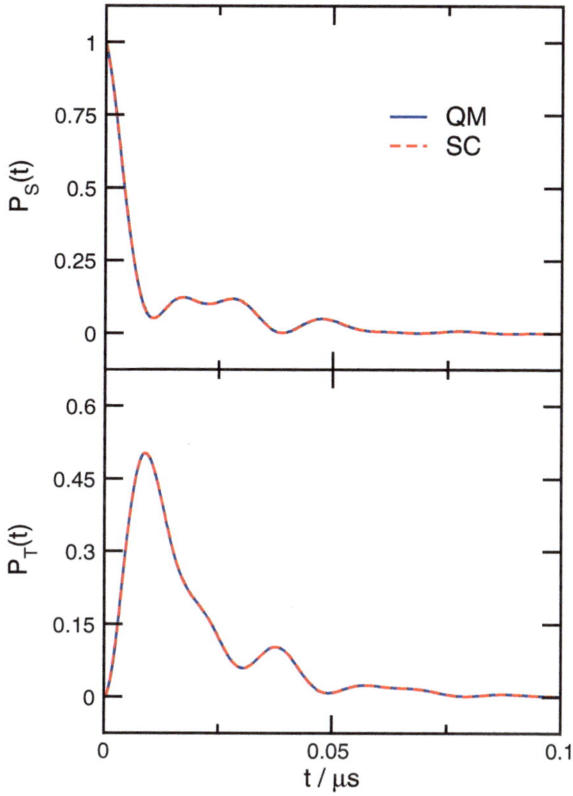

Fig. C.1 Comparison of the quantum mechanical (QM) and semiclassical (SC) singlet and triplet probabilities of a radical pair with $\bar{\omega}_1 = \omega_1 = (0, 0, 1)$, $\bar{\omega}_2 = \omega_2 = -(\sqrt{1/2}, \sqrt{1/3}, \sqrt{1/5})$, $k_S = \sqrt{1/7}$, and $k_T = \sqrt{1/11}$. The electron spin coupling tensor $\mathbf{D}'$ is defined in Eq. (C.14)

when the coupling between the two electron spins in the radical pair is accounted for. This is demonstrated by Fig. C.1, which shows the singlet and triplet probability of a radical pair with no nuclear spins, asymmetric recombination rates, $\omega_1 \neq \omega_2$, and an anisotropic electron spin coupling tensor

$$\mathbf{D}' = \begin{pmatrix} -0.9999847 & -0.7369246 & 0.5112104 \\ -0.7369246 & -0.0826997 & 0.0655341 \\ 0.5112104 & 0.0655341 & -0.5620818 \end{pmatrix}. \tag{C.14}$$

Figure C.1 is entirely analogous to Fig. 3.1, and shows that when the terms in Eqs. (C.1)–(C.3) are added to Eqs. (3.4)–(3.6), the equations of motion remain exact in the limit of no nuclear spins. Note that when the electron spin coupling tensor is non-zero, Eqs. (C.1)–(C.3) couple the equations of motion of the one- and two-electron classical spin variables, and so the singlet and triplet probabilities depend on all 16 classical variables.

Appendix D
Schulten–Wolynes Expressions

The closed form expression for the ensemble average of the singlet probability of a singlet-born radical pair within Schulten–Wolynes theory is [1]

$$
\begin{aligned}
\mathrm{P_S}(t) = \frac{1}{4} + \frac{1}{4\tau_1^3\tau_2^3\omega^6}\Big[& \Big\{\tau_1\omega\left(\tau_1^2\omega^2 + 4e^{-t^2/\tau_1^2}\cos\omega t - 4\right) + 2\sqrt{\pi}e^{-\tau_1^2\omega^2/4}y(\tau_1)\Big\}\times \\
& \Big\{\tau_2\omega\left(\tau_2^2\omega^2 + 4e^{-t^2/\tau_2^2}\cos\omega t - 4\right) + 2\sqrt{\pi}e^{-\tau_2^2\omega^2/4}y(\tau_2)\Big\}\Big] + \\
& \frac{1}{2\tau_1^2\tau_2^2\omega^4}\Big[e^{-t/\tau_1^2}e^{-t/\tau_2^2}\Big\{2\omega t\cos\omega t + \left(\tau_1^2\omega^2 - 2\right)\sin\omega t\Big\}\times \\
& \Big\{2\omega t\cos\omega t + \left(\tau_2^2\omega^2 - 2\right)\sin\omega t\Big\}\Big] + \\
& \frac{1}{2\tau_1^2\tau_2^2\omega^4}\Big[\Big\{z(\tau_1) - \sqrt{\pi}e^{-\tau_1^2\omega^2/4}y(\tau_1)\Big\}\times\Big\{z(\tau_2) - \sqrt{\pi}e^{-\tau_2^2\omega^2/4}y(\tau_2)\Big\}\Big],
\end{aligned}
\tag{D.1}
$$

where

$$
\begin{aligned}
y(x) &= -i\,\mathrm{erf}\left(\frac{t}{x} - \frac{ix\omega}{2}\right) + i\,\mathrm{erf}\left(\frac{t}{x} + \frac{ix\omega}{2}\right) - 2i\,\mathrm{erf}\left(\frac{ix\omega}{2}\right), \\
z(x) &= x\omega\left(e^{-t^2/x^2}\left(\left(x^2\omega^2 - 2\right)\cos\omega t - 2\omega t\sin\omega t\right) + 2\right).
\end{aligned}
\tag{D.2}
$$

The Schulten–Wolynes approximations to the xx, xy, and zz components of the spin correlation tensor are [2]

A. Lewis, *Spin Dynamics in Radical Pairs*, Springer Theses,
https://doi.org/10.1007/978-3-030-00686-0

$$R_{xx}^{(i)}(t) = \frac{1}{2\omega^2\tau_i^2}\left[e^{-t^2/\tau_i^2}\left((\omega^2\tau_i^2 - 2)\cos\omega t - 2\omega t \sin\omega t\right) + 2 - \frac{\sqrt{\pi}}{\omega\tau_i}e^{-\omega^2\tau_i^2/4}y(\tau_i)\right],$$

$$R_{xy}^{(i)}(t) = \frac{e^{-t^2/\tau_i^2}}{2\omega^2\tau_i^2}\left[2\omega t\cos\omega t + \left(\tau_i^2\omega^2 - 2\right)\sin\omega t\right],$$

$$R_{zz}^{(i)}(t) = \frac{1}{2\omega^2\tau_i^2}\left[\left(\omega^2\tau_i^2 + 4e^{-t^2/\tau_i^2}\cos\omega t - 4\right) + 2\frac{\sqrt{\pi}}{\omega\tau_i}e^{-\omega^2\tau_i^2/4}y(\tau_i)\right]. \tag{D.3}$$

When the hyperfine interactions in a radical pair are isotropic, these are the only components which need to be specified, since

$$\begin{aligned} R_{xx}^{(i)}(t) &= R_{yy}^{(i)}(t), \\ R_{xy}^{(i)}(t) &= -R_{yx}^{(i)}(t), \\ R_{xz}^{(i)}(t) = R_{zx}^{(i)}(t) &= R_{yz}^{(i)}(t) = R_{zy}^{(i)}(t) = 0. \end{aligned} \tag{D.4}$$

In Eqs. (D.1) and (D.3), τ_i describes the distribution of the resultant of the nuclear spin vectors, and is defined in Eq. (3.29), $\omega = -\gamma_e B$ is Larmor frequency of the electrons in the radical pair, and $\mathrm{erf}(x)$ is the error function,

$$\mathrm{erf}(x) = \frac{2}{\sqrt{\pi}}\int_0^x e^{-y^2}\mathrm{d}y. \tag{D.5}$$

References

1. Rodgers, C. T. (2007). *Magnetic Field Effects in Chemical Systems*. Ph.D. thesis, University of Oxford.
2. Manolopoulos, D. E. & Hore, P. J. (2013). An improved semiclassical theory of radical pair recombination reactions. *Journal of Chemical Physics 139*, 124106.

Appendix E
The Hyperfine Interactions of the Cryptochrome Radical Pair

In Tables E.1 and E.2 we list the anisotropic hyperfine coupling tensors of the flavin adenide nucleotide radical ($FAD^{\bullet -}$) and tryptophan radical ($Trp_C^{\bullet +}$) respectively. These form the cryptochrome-based radical pair described in Sect. 5.1.2 thought to be responsible for avian magnetoreception. The hyperfine tensors were calculated using B3LYP density functional theory[1, 2] and the EPR-II basis set [3]. Figures E.1 and E.2 show positions of the magnetic nuclei in the FAD and Trp_C radical which correspond to the hyperfine coupling tensors listed in Tables E.1 and E.2.

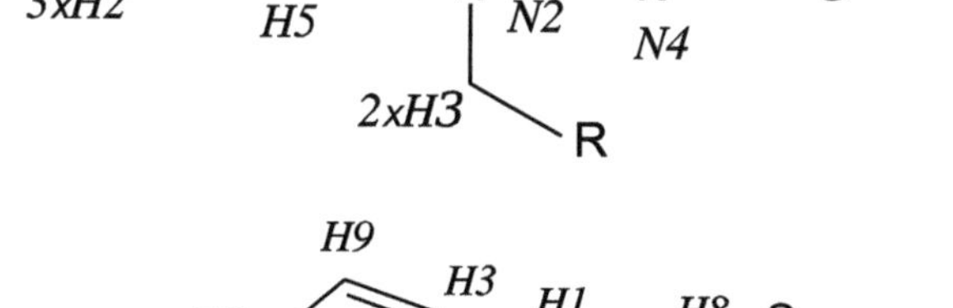

Fig. E.1 The FAD cofactor found in cryptochrome with the magnetic nuclei labelled

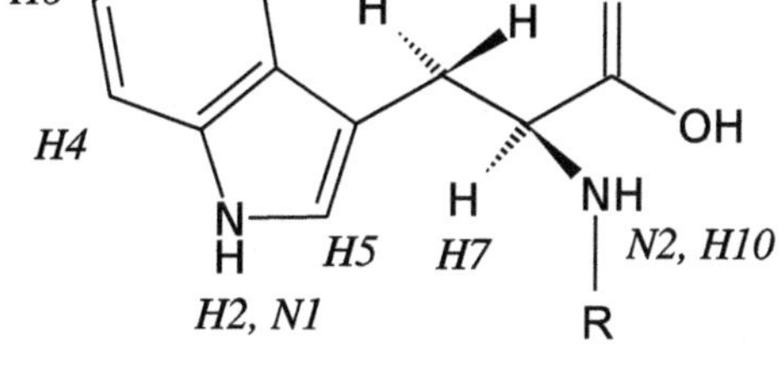

Fig. E.2 The Trp_C residue found in cryptochrome with the magnetic nuclei labelled

A. Lewis, *Spin Dynamics in Radical Pairs*, Springer Theses,
https://doi.org/10.1007/978-3-030-00686-0

Table E.1 The hyperfine coupling tensors of the $FAD^{\bullet -}$ radical in mT, calculated using B3LYP density functional theory and the EPR-II basis set. The components of the tensors are listed in the order $pq = xx, xy, xz, yx, \cdots, zz$

Nucleus	A'_{pq}	Nucleus	A'_{pq}	Nucleus	A'_{pq}	Nucleus	A'_{pq}
N1	−0.098853074	N2	−0.0189666726	H1	−0.2568678491	H2	0.4399413333
	0.0038837628		−0.0048409688		−0.1273243546		0
	0		0		0		0
	0.0038837628		−0.0048409688		−0.1273243546		0
	−0.0880561739		−0.0195782243		−0.4711376546		0.4399413333
	0		0		0		0
	0		0		0		0
	0		0		0		0
	1.7569		0.60458		−0.4336		0.4399413333
H3	0.407031	H4	−0.1415756667	H5	0.0554170121	H6	−0.0146225342
	0		0		0.0056842845		0.0286310253
	0		0		0		0
	0		0		0.0056842845		0.0286310253
	0.407031		−0.1415756667		0.1191100193		0.0226151305
	0		0		0		0
	0		0		0		0
	0		0		0		0
	0.407031		−0.1415756667		−0.004986		−0.064554
N3	−0.0403621991	N4	−0.0224869399				
	0.0045002445		0.0025191054				
	0		0				
	0.0045002445		0.0025191054				
	−0.0357786575		−0.0259267146				
	0		0				
	0		0				
	0		0				
	−0.038819		0.038045				

Table E.2 The hyperfine coupling tensors of the $Trp_C^{\bullet +}$ radical in mT, calculated using B3LYP density functional theory and the EPR-II basis set. The components of the tensors are listed in the order $pq = xx, xy, xz, yx, \cdots, zz$

Nucleus	A'_{pq}	Nucleus	A'_{pq}	Nucleus	A'_{pq}	Nucleus	A'_{pq}
H1	1.5808074002	H2	−0.9920313196	N1	−0.0336355053	H3	−0.5596032693
	−0.045287774		−0.2091239528		0.0924456014		−0.1956258627
	−0.0505827536		−0.2003408046		−0.1353497899		−0.1657438187
	−0.045287774		−0.2091239528		0.0924456014		−0.1956258627
	1.5574613327		−0.2631022871		0.3303312815		−0.4020115833
	0.0988262033		0.2802929192		−0.5318203648		0.076235815
	−0.0505827536		−0.2003408046		−0.1353497899		−0.1657438187
	0.0988262033		0.2802929192		−0.5318203648		0.076235815
	1.6752075037		−0.539817385		0.6679701897		−0.5021472426
H4	−0.4354610485	H5	−0.2842866829	H6	−0.05059919	H7	−0.0601236697
	−0.1541117685		0.1757406004		0.0622015673		0.0036893058
	−0.123872542		0.1525438652		0.0889307955		0.0330896062
	−0.1541117685		0.1757406004		0.0622015673		0.0036893058
	−0.277686742		−0.2797937779		−0.3100128398		−0.0251024367
	0.08637812		0.0974897683		−0.0296595099		0.0110551878
	−0.123872542		0.1525438652		0.0889307955		0.0330896062
	0.08637812		0.0974897683		−0.0296595099		0.0110551878
	−0.377783436		−0.2699373143		−0.264226336		−0.1939736323
N2	0.1294731476	H8	0.1633847872	H9	0.005082777	H10	0.0328654946
	−0.0133833225		−0.0229751692		0.0615701138		−0.0321038673
	0.0075048796		−0.0063992933		0.0694326568		0.0587459026
	−0.0133833225		−0.0229751692		0.0615701138		−0.0321038673
	0.1728840159		−0.0081074178		−0.0665195585		0.0253328922
	−0.0249044694		0.0157763205		0.0390812016		−0.0168505051
	0.0075048796		−0.0063992933		0.0694326568		0.0587459026
	−0.0249044694		0.0157763205		0.0390812016		−0.0168505051
	0.1370568271		−0.0181895625		−0.0586105041		0.0117348028

References

1. Lee, C., Yang, W. & Parr, R. G. (1988). Development of the Colle-Salvetti correlation-energy formula into a functional of the electron density. *Physical Review B 37*, 785–789.
2. Becke, A. D. (1993). A new mixing of Hartree-Fock and local density-functional theories. *The Journal of Chemical Physics 98*, 1372.
3. Barone, V. (1995). *Recent advances in density functional methods: Part I*. Singapore: World Scientific.

MIX
Papier aus verantwortungsvollen Quellen
Paper from responsible sources
FSC® C105338

If you have any concerns about our products,
you can contact us on
ProductSafety@springernature.com

In case Publisher is established outside the EU,
the EU authorized representative is:
Springer Nature Customer Service Center GmbH
Europaplatz 3, 69115 Heidelberg, Germany

Printed by Libri Plureos GmbH
in Hamburg, Germany